Differential Diagnosis in
Neurology
and
Neurosurgery

Differential Diagnosis in
Neurology
and
Neurosurgery

Virendra Deo Sinha
Professor and Head
Department of Neurosurgery
SMS Medical College, Jaipur

Jitin Bajaj
Fellow, Epilepsy and Functional Neurosurgery
All India Institute of Medical Sciences, New Delhi

Trilochan Srivastava
Professor
Department of Neurosurgery
SMS Medical College, Jaipur

CBS

CBS Publishers & Distributors Pvt Ltd

New Delhi • Bengaluru • Chennai • Kochi • Kolkata • Mumbai
Bhopal • Bhubaneswar • Hyderabad • Jharkhand • Nagpur • Patna • Pune
Uttarakhand • Dhaka (Bangladesh) • Kathmandu (Nepal)

Differential Diagnosis in
Neurology
and
Neurosurgery

ISBN: 978-93-87742-81-9

Copyright © Authors and Publisher

First Edition: 2018

Reprint: 2020

Published by Satish Kumar Jain and produced by Varun Jain for

CBS Publishers & Distributors Pvt Ltd

4819/XI Prahlad Street, 24 Ansari Road, Daryaganj, New Delhi 110 002, India.
Ph: 011-23289259, 23266861, 23266867   Website: www.cbspd.com
Fax: 011-23243014   e-mail: delhi@cbspd.com; cbspubs@airtelmail.in.
Corporate Office: 204 FIE, Industrial Area, Patparganj, Delhi 110 092

Ph: 011-4934 4934   Fax: 011-4934 4935   e-mail: publishing@cbspd.com; publicity@cbspd.com

Branches

- **Bengaluru:** Seema House 2975, 17th Cross, K.R. Road,
  Banasankari 2nd Stage, Bengaluru 560 070, Karnataka
  Ph: +91-80-26771678/79   Fax: +91-80-26771680   e-mail: bangalore@cbspd.com
- **Chennai:** 7, Subbaraya Street, Shenoy Nagar, Chennai 600 030, Tamil Nadu
  Ph: +91-44-26680620, 26681266   Fax: +91-44-42032115   e-mail: chennai@cbspd.com
- **Kochi:** 68/1534, 35, 36, Power House Road, Opp. KSEB, Kochi 682018, Kerala
  Ph: +91-484-4059061-65   Fax: +91-484-4059065   e-mail: kochi@cbspd.com
- **Kolkata:** 6/B, Ground Floor, Rameswar Shaw Road, Kolkata-700 014, West Bengal
  Ph: +91-33-22891126, 22891127, 22891128   e-mail: kolkata@cbspd.com
- **Mumbai:** 83-C, Dr E Moses Road, Worli, Mumbai-400018, Maharashtra
  Ph: +91-22-24902340/41   Fax: +91-22-24902342   e-mail: mumbai@cbspd.com

Representatives

• **Bhopal**	0-8319310552	• **Bhubaneswar**	0-9911037372
• **Hyderabad**	0-9885175004	• **Jharkhand**	0-9811541605
• **Nagpur**	0-9421945513	• **Patna**	0-9334159340
• **Pune**	0-9623451994	• **Uttarakhand**	0-9716462459
• **Dhaka (Bangladesh)**	01912-003485	• **Kathmandu (Nepal)**	977-9818742655

Printed at Rashtriya Printer, Dilshad Garden, Delhi, India

Preface

Neurological and neurosurgical disorders remain one of the most intriguing cases in the field of medicine. Even in the current era, history taking and examination are the most useful and cost-effective techniques to select an investigation, and reach to a definitive diagnosis. While there are many eminent books, which give direct knowledge about the diseases, there is lack of a good book, which starts from the patient's complaint and reach to a diagnosis. This book has been written taking in view of approach to a neurological/neurosurgical patient. In each chapter, a common complaint is chosen and relevant history taking and examination is told. Relevant investigations and differential diagnosis are also given. Every chapter starts with an introduction and a flow diagram to give a quick view.

To the best of our belief, this book will be highly useful to residents of neurology, neurosurgery, graduate and undergraduate students of medicine, and physicians at large.

Virendra Deo Sinha
Jitin Bajaj
Trilochan Srivastava

Contents

Preface   *v*

1. Headache   **1**

2. Seizures   **24**

3. Speech   **34**

4. Vomiting   **40**

5. Anosmia   **44**

6. Visual Disturbance   **51**

7. Diplopia   **60**

8. Ptosis   **67**

9. Proptosis   **75**

10. Facial Pain   **81**

11. Facial Asymmetry   **90**

12. Taste   **100**

13. Hearing Loss   **106**

14. Tinnitus   **113**

15. Vertigo   **119**

16. Nystagmus   **126**

17. Dysphagia   **133**

18. Neck Pain   **141**

19. Low Back Pain   **149**

20. Sensory Impairment   **157**

21. Weakness   **165**

22. Movement Disorders   **186**

23. Gait   **201**

24. Neurogenic Bladder   **210**

Index   *219*

Headache

Headache

History

- Age
- Sex
- Onset
  - Sudden
  - Insidious
- Duration
- Progression
- Location
- Radiation
- Nature
- Intensity
- Frequency
- Diurnal variation
- Aggravating factors
- Relieving factors
- Trauma
- Associated symptoms
- Past history
- Personal history
- Family history
- Treatment taken and present status
- Anything else

Examination

General

- Vitals
- Consciousness
- Ocular
- Skull
- Scalp
- Paranasal sinuses
- Teeth
- Nose
- Ears

Neurologic

- MMSE
- Higher mental functions
- Ocular
- Sensorimotor
- Signs of meningitis

Definition: Pain between the orbits and occiput.

First step in evaluation of a headache is to classify whether it is **primary headache** (like migraine, tension type, cluster) or **secondary headache** due to either a non-neurologic cause (hypertension, sinusitis, odantalgia, depression), or a neurologic cause (like raised ICP, tumour, subarachnoid hemorrhage, trigeminal neuralgia, etc.).[1]

History

- **Age and sex:** Migraine is a disease of adolescent females which can progress up to adulthood. Children mostly have refractive errors. Most common cause of pediatric headache in emergency settings is viral illness.[2] Vomiting with headache in a child may be because of posterior fossa mass. Adults mostly have hypertension as their cause of headache. Elderly patients with severe headache of short duration may point towards temporal arteritis.

- **Onset:** Sudden onset includes **vascular causes**—subarachnoid hemorrhage (SAH), arterial dissection, cerebrovascular accident (CVA), reversible cerebral vasoconstriction syndrome (RCVS), **sudden rise in intracranial pressure** (ICP) like tumour bleed, **pituitary apoplexy**, and **acute hypertensive crisis**. Insidious onset ones includes others (as mentioned below).

- **Duration:** Question should be asked: How long do you have headache?

  Short duration: It includes sinusitis, glaucoma, post-traumatic, retrobulbar neuritis, drugs/toxins (vasodilators), infection like meningitis or encephalitis, hydrocephalus, RCVS.

- **Intermediate:** Hydrocephalus, ICSOL, infections like tuberculosis, abscess, chronic meningitis, temporal arteritis, Intracranial hypertension/hypotension

  Long: Tension type headache, migraine, cluster headache, refractory errors, cervical spondylosis, transformed (after long period of migraine there can occur a mix of migraine + tension headaches), medication overuse headache, drugs (vasodilators)

- **Progression:** Progressive headache is suggestive of organic cause.

- **Location:** Though location is not strictly linked to the cause, but it points towards etiology. It may be **bifrontal** (refractive errors, raised ICP, frontal sinusitis), **holocranial** (tension, raised ICP), **vertex** (psychotic), **band like/ occipital** (tension), **hemicranial** (migraine/cluster), **temporal** (migraine, temporal arteritis), **periorbital** (eye diseases like glaucoma, iridocyclitis, cavernous sinus lesions, cluster), **localized over one area** (tumours like meningioma), **neck, shoulder or arm pain** (cervicogenic). Bilateral headache may be because of hypertension.

  Raised ICP headache are bifrontal because of irritation of dura by the expanding frontal horns.

- **Radiating/nonradiating:** Tension headache may be cervicogenic in origin arising from neck and jaw muscles, and often get relieved with suboccipital massage. Migraine headache originates unilaterally, and may become generalized. Facial pain like trigeminal neuralgia may radiate to one side of head (*see* chapter of facial pain).

- **Nature: Throbbing/pulsatile** (migraine, hypertension, intracranial arteriovenous malformations, anemia and aneurysm), **bursting** (raised ICP), **dull aching** (mass lesion), **tightness/pressure/band like** (tension), **sharp lancinating** (trigeminal neuralgia), **boring pain** (skull diseases).

- **Intensity:** Patients should be asked to grade their pain on visual analog scale from **0 to 10**, where 10 means most severe. It should be noted that it is subjective and does not tell whether it is primary or secondary headache, but it definitely helps in better follow-up. Increasing severity points towards a secondary cause of headache.

- **Frequency:** Patients should be asked to maintain a headache diary. Enquiry should be done whether it is daily, weekly or monthly.[3] Increasing frequency points toward a secondary cause. Fixed seasonal pattern occurs in cluster headache and cyclic pattern occurs in cyclic migraine. Recurrent thunderclap headache is a hallmark of RCVS, though it may be absent also.[4]

- **Diurnal variation: Early morning or headache on awakening** includes causes due to raised ICP, migraine, decreased sleep (includes obstructive sleep apnoea), uraemia, sinusitis, cervical spondylosis. **Evening** headache includes cluster, tension, arteritis, and stress. Headaches more likely to present at **night** are hypnic headache syndrome, cluster headache, SUNCT syndrome or exploding head syndrome,[3] tumours and meningitis.

- **Aggravating factors:** Bending forward (sinusitis), coughing/ sneezing, straining/worsening with head position (raised ICP), anger/excitement/irritation (migraine/tension), worsened by bright light/noise (migraine), menstrual disturbances (migraine). Headache due to rapid descent in height (aeroplane, car) may occur due to imbalance between intrasinusal pressure and external air pressure,[5] hunger/ food consumption/drink—migraine/cluster.

- **Relieving factors:** Darkness, sleep, and vomiting relieves headache due to migraine. Vomiting also reduces the headache due to raised ICP because of two reasons: Firstly it makes the patient to sit or stand from lying down position which reduces the venous pressure, and secondly it causes exhalation of lot of carbon dioxide which causes vasoconstriction of the arteries leading to reduced ICP. Headache relieving on sitting from lying down also points towards raised ICP, and reverse is seen in low pressure headache like that after lumbar puncture.

- **Trauma:** Headache post-trauma can be because of **head trauma** like skull fractures, concussion, space occupying hematomas, dural irritation, CSF leaks, or **neck trauma** like vascular dissection.

- **Associated symptoms:** Seizures, syncope, double vision should be asked to rule out secondary causes. Transient loss of vision with sudden change in posture/or increased headache is seen in raised ICP. Fever, chills, neck rigidity may be seen in meningitis. Hypersensitivity phenomenon like photophobia, phonophobia, osmophobia are seen in migraine. Myalgias are seen in tension/viral infection. Sudden repeated loss of consciousness may be seen in colloid cyst; while repeated syncope may be seen in migraine/hysteria. Rhinorrhoea/lacrimation/redness of eyes are seen in trigeminal autonomic cephalalgias like cluster headache. Trigger points are seen in trigeminal neuralgia. Palpitations are seen in cardiovascular cause. Tender painful superficial temporal artery is seen in temporal arteritis. Aura may be seen both in migraine or a seizure, with the difference in duration of it, which is short in seizure and long in migraine (20–30 min).

- **Past history:** Past systemic malignancies may metastasize to brain. Hypertension, coagulation disorders, etc. can influence the cause. A patient of bronchial asthma has contraindications for prescribing propanolol for migraine.
- **Personal history:** Addiction history should be taken. Alcohol can precipitate cluster headache. Sleep problems should be asked. Ask about the impact of headache on patients life—this will tell us how severe the headache is and helps in differentiating out the primary and secondary causes.
- **Family history:** Migraine may run in the family.
- **Present status and treatment taken:** This history is very important. A patient of unilateral headache and facial pain responding to carbamazepine fits into the category of trigeminal neuralgia. Many patients have a previous scan done before coming to us.
- **Anything else?[3]:** Sometimes, patients reveal other important information on asking this question. He may reveal galactorrhoea, recent malarial fever, HIV infections, feuds in the family, etc.

Examination

General Examination

- Vital parameters: Raised blood pressure may be seen in hypertensive headaches and raised ICP. Fever may be seen in meningitis, encephalitis, or systemic infections.
- See for consciousness, orientation, confusion, behavior, irritability, mood, flight of ideas, etc.
- Skull: One should look for signs of raised ICP—like in a child—serial increased head circumference (measured from widest part of forehead above eyebrows to the occipital protruberence), sunset eyes (upgaze paresis with eyes driven downwards), bulging fontanels, Macewen or crack pot sign (on percussion a resonant sound is produced due to split sutures). Unshaved face may be seen in trigeminal neuralgia, teeth and scalp should be examined for local causes; paranasal sinuses should be palpated and percussed to see for any tenderness.

Neurological Examination

- Minimental state examination:[6] Originally this score was developed for dementia, but because of its fast and objective nature it can be applied to screen the patients for any disturbed higher mental functions. It includes a set of questionnaire consisting of maximum score of 30. A score of less than 24 is set to be abnormal in an educated person. The test is not timed but generally can be finished in 5–10 min. It is given in Table 1.1.

Maximum score	Score	Orientation
		Table 1.1: Minimental state examination
5	( )	What is the (year) (season) (date) (day) (month)?
5	( )	Where are we: (state) (county) (town) (hospital) (floor)
		Registration
3	( )	Name 3 objects: 1 second to say each. Then ask the patient all 3 after you have said them. Give 1 point for each correct answer. Then repeat them until he learns all 3. Count trials and record.
		Attention and Calculation
5	( )	Serial 7's. 1 point for each correct. Stop after 5 answers. Alternatively spell "world" backwards.
		Recall
3	( )	Ask for the 3 objects repeated above. Give 1 point for each correct.
		Language
9	( )	Name a pencil, and watch (2 points) Repeat the following "No ifs, ands or buts" (1 point) Follow a 3-stage command:" Take a paper in your right hand, fold it in half, and put it on the floor" (3 points) Read and obey the following: Close Your Eyes (1 point) Write a sentence (1 point) Copy a design (1 point)
		Total Score

For uneducated and rural population, a hindi version of MMSE has been developed by Indo-US Cross National Dementia Epidemiological Study.[7] It has modified questions as given in Table 1.2.

Table 1.2: Hindi minimental state examination		
Maximum score	*Score*	*Orientation*
5	( )	What is the (time of day) (day of week) (date) (month) (season)?
5	( )	Where are we: (district) (post office) (village) (block) (which place is this/whose house is this)
		Registration
3	( )	Tell him that I have brought 3 objects from Delhi— "a mango, a chair, and a coin". Then ask the patient all 3 after you have said them. Give 1 point for each correct answer. Then repeat them until he learns all 3. Count trials and record.
		Attention and Calculation
5	( )	Serial 7's. 1 point for each correct. Ask the week days backwards by giving examples. Stop after 5 answers (omit Sunday and Saturday for scoring). For calculation, give example that "A man has 20 rupees for bus fare. Every day, he spends 3 rupees on his bus fare. After spending the first day's bus fare, he will be left with 17 rupees. How much money will be left after the next day's bus fare ... and the next day's bus fare ..." The first five consecutive responses are scored.
		Recall
3	( )	Ask for the 3 objects repeated above (mango, chair and coin), by asking "What are the three objects I told you I brought with me from Delhi?". Give 1 point for each correct.
		Language
9	( )	Name a pen, and watch (2 points) Repeat the following "ना, ना ही ये, ना ही वो" (1 point) Follow a 3-stage command: "Take a paper in your right hand, fold it in half, and put it on the floor" (3 points)

Contd.

		Table 1.2: Hindi minimental state examination (Contd.)
Maximum score	*Score*	*Orientation*
		Read and obey the following: Examiner says "Look at me and do exactly what I do" and then closes his eyes for 3 seconds (follow example) while the co-examiner observes and record the patient's responses. (1 point) Ask the patient: "Tell me something about your house". Award 1 point to any complete sentence offered in response. Copy a design: A diamond within a square (1 point)
		Total Score

Detailed Higher Mental Function Examination[8]

It is done when MMSE comes out to be abnormal.

- Level of consciousness: See and classify whether patient is in *normal, clouded, delirium, obtunded, stupor, or coma.*
- Appearance and general behaviour: Observe the patient's physical appearance (apparent vs. stated age), grooming (immaculate/unkempt), dress (subdued/riotous), posture (erect/kyphotic), and eye contact (direct/furtive).
- Speech and motor activity: Listen to the spontaneous speech of the patient. Overall, motor activity of the patient should also be noted.
- Affect and mood: Affect is the immediate emotional expression, and mood is the more sustained emotional makeup. Both of them can be classified as dysphoric (depression, anxiety, guilt), euthymic (normal), or euphoric (implying a pathologically elevated sense of well-being).
- Thought and perception: Ask: "Have you ever seen or heard things that other people could not see or hear? Have you ever seen or heard things that later turned out not to be there?" Patient may be having *delusional thinking* (a fixed, false belief), *hallucinations* (false sensory perceptions without real stimuli), or *illusions* (misperceptions of real stimuli).
- Attitude and insight: See for patient's attitude toward the examiner, other individuals, or his illness. He may have a sense of hostility, anger, helplessness, pessimism, overdra-

matization, self-centredness, or passivity. Also see for his attitude towards the illness.

- Examiner's reaction to the patient: Examiner may feel dysphoric with a depressed patient, frustrated with a help-rejecting one, and off-touch with a schizophrenic patient.
- Attention and vigilance: Attention can be tested by serial subtraction test. To test vigilance: Ask the patient to tap whenever a letter or number comes in your example. For example, he can tap whenever "A" comes in "G, H, K, X, L, N, A, O, P".
- Language: It includes assessment of spontaneous speech, comprehension of spoken commands, reading ability, reading comprehension, writing, and repetition. In the *spontaneous speech* look for phonemic errors (disturbed pronunciation), semantic errors (disturbed meaning of words), and neologisms (meaningless non-words that have a specific meaning for the patient). *Comprehension* is tested by first asking simple 'yes or no' questions like: Do you take off your shoes while going to your religious place? Then gesture alone questions are asked like—"point in the room where people sit". After this, he is asked for a complex motoric response like "give this pen to me". *Repetition* is asked in English as "*No ifs, and or buts*", and in Hindi as "ना, ना ही ये, ना ही वो". *Word finding difficulty* is assessed by asking him to make all words starting with a letter like 'S' *Reading* is tested by asking him to read short sentences. Reading comprehension can also be tested by asking him to read "Close your eyes" and watching his response. *Writing* can be tested by asking him to write his name.
- Memory: It includes immediate, short-term, and long-term memory. *Immediate memory* is tested by repeating digit spans both forward and backwards. Normal forward digit span is 7 ± 2 and backward digit span is 5 ± 2. *Short-term memory* is tested by having the patient learn 4 different objects and recalling them after 3–5 minutes. Similarly, visual memory can also be tested by hiding 4 different objects and asking him to recall him where the objects were hidden. *Long-term memory* can be tested by asking him to recall his place of birth.

Constructional ability and praxis: Apraxia is the inability to do a learnt act voluntarily in absence of motor weakness. See for ideomotor apraxia (ability to initiate or manipulate the object, but inability to pretend a given action-by asking the patient to sew with an imaginary needle). See for ideational apraxia—by asking him to do a 3-stage command. Constructional ability is tested by asking the patient to draw a clock with a specific timing, for example, 9:00.

- Abstract reasoning: Ask the patient to explain proverbs.

Lobar Examination

A. **Frontal lobe:** Frontal assessment battery (FAB) testing[9] should be done. It has a maximum score of 18, and higher scores indicate better performance.

   i. Frontal assessment battery test[9]

      1. Similarities (Conceptualization): In what way are they alike?

        – A banana and an orange

        – A table and a chair

        – A tulip, a rose, and a daisy

      Score (only category responses—fruits, furnitures, flowers—are considered correct)

      2. Lexical fluency (mental flexibility)

      "Say as many words as you can beginning with letter 'S', any word except surnames or proper nouns". If the patient gives no response during the first 5 seconds, then give example, say: "For instance, snake." If the patient pauses 10 seconds, stimulate him by saying: "any word beginning with the letter 'S'. The time allowed is 60 seconds."

      Score: > 9 words: 3, 6–9 words: 2, 3–5 words: 1, <3 words: 0

      3. Motor series "Luria test" (programming)

      Examiner says: "Look carefully at what I'm doing." Show the patient with your left hand alone three times the series of fist-edge-palm.

      Now say to the patient: "Now, with your right hand do the same series, first with me, then alone."

      Perform the series three times with the patient, then say to him/her:

      "Now, do it on your own."

Score

Patient performs six correct consecutive series alone: 3

Patient performs at least three correct consecutive series alone: 2

Patient fails alone, but performs three correct consecutive series with the examiner: 1

Patient cannot perform three correct consecutive series even with the examiner: 0

4. Conflicting instructions (sensitivity to interference)

Examiner says: "Tap twice when I tap once."

To ensure that the patient has understood the instruction, a series of 3 trials is run: 1-1-1.

Then say: "Tap once when I tap twice."

To ensure that the patient has understood the instruction, a series of 3 trials is run: 2-2-2.

The examiner then performs the following series: 1-1-2-1-2-2-2-1-1-2.

Score No errors: 3, 1–2 errors: 2, > 2 errors: 1

Patient taps like the examiner at least four consecutive times: 0

5. Go-No-Go (inhibitory control)

Examiner says: "Tap once when I tap once."

To ensure that the patient has understood the instruction, a series of 3 trials is run: 1-1-1.

Then say: "Do not tap when I tap twice."

To ensure that the patient has understood the instruction, a series of 3 trials is run: 2-2-2.

The examiner then performs the following series: 1-1-2-1-2-2-2-1-1-2.

Score No errors: 3, 1–2 errors: 2, > 2 errors: 1

Patient taps like the examiner at least four consecutive times: 0

6. Prehension behaviour (environmental autonomy)

Say to the patient: "Do not take my hands."

Sit in front of the patient. Place the patient's hands palm up on his knees. Without saying anything or looking at the patient, bring your hands close to the patient's hands and touch the palms of both the patient's hands, to see if he will spontaneously take them. If the patient takes your hands, try again after asking the patient: "Now, do not take my hands."

Score

Patient does not take the examiner's hands: 3

Patient hesitates and asks what he/she has to do: 2

Patient takes the hands without hesitation: 1

Patient takes the examiner's hand even after he/she has been told not to do so: 0

Interpreting results: A cutoff score of 12 on the FAB has a sensitivity of 77% and specificity of 87% in differentiating between frontal dysexecutive type dementias and DAT.

ii. **Perseveration:** Give a drawing to be made containing of alternating parts, for example

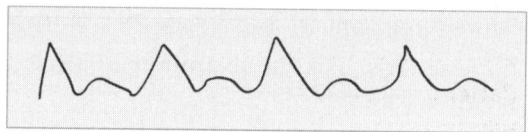

If abnormal, then patient may make it wrong like this:

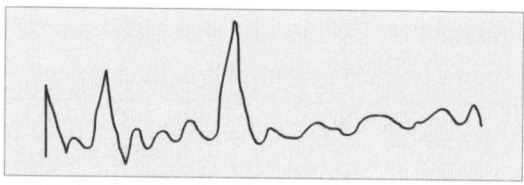

iii. **Fluency:** Examine for fluency to search for Broca's aphasia (as explained earlier)

iv. **See for social inhibition, urinary incontinence, lack of attention broad-based gait.**

B. **Parietal lobe:**

   i. Agnosia: Inability to identify objects by tactile sensations (explained in sensory loss chapter).

   ii. Apraxia: Inability to perform learnt movements in absence of weakness.

iii. Constructional apraxia and hemineglect: Ask patient to draw a clock. He may not draw left-sided numbers (seen in right parietal lobe).

iv. Geographical disorientation (right parietal lobe): Ask him with eyes closed where is the door.

v. Testing for Gerstman's syndrome (found in dominant temporal lobe lesions). Consists of
 – Acalculia
 – Right left disorientation
 – Finger agnosia
 – Agraphia

C. Temporal lobe:

i. Comprehension: For Wernicke's aphasia (as explained earlier).

ii. Visual memory: For right temporal lobe (as explained earlier).

iii. Verbal memory: For left temporal lobe (similar to examining of short-term memory examination)

iv. If bilateral temporal lobes: Patient may show features of Klüver-Bucy syndrome like oversexuality, hyperorality, visual agnosia, docility.

D. Occipital lobe:

i. Visual fields: See chapter on visual loss

ii. Naming of objects

iii. Naming of colours

iv. Recognition of faces

v. Denial of blindness (Anton's syndrome): Patient may deny blindness.

- **Ocular examination:** Acuity (refractive errors—better explained in visual loss chapter), tenderness (inflammations), nystagmus (for localization—better explained in nystagmus chapter), diplopia, papilledema or optic atrophy (explained in visual loss chapter).

- **Sensorimotor examination:** Examine for sensory and motor functions (explained in respective chapters). See for any neurological deficits in them.

- **Signs of meningitis.** Inability to flex the neck. *Brudzinski's sign:* Put the patient in supine position.

On flexing the neck there will be involuntary lifting of legs. *Kernig's sign*: In the supine position with the patient's hip and knee flexed 90 degrees, try to extend the knee—this will produce pain and resistance.

Investigations: Appropriate investigations are needed when either of the red flag signs, remembered by mnemonic **SNOOP**[10] is fulfilled. These include:

S: Systemic signs like fever, myalgias, weight loss, HIV infection, known of malignancy

N: Neurological signs of symptoms like confusion, focal neurological deficits.

O: Onset—acute, sudden, thunderclap

O: Onset after age 40

P: Pattern change—progressive headache with loss of headache—free periods, change in type of headache. Headache in children <5 years old.[11]

Choosing Wisely Campaign

American Board of Internal Medicine Foundation and the American Headache society (*http://www.choosingwisely.org/ societies/american-headache-society/*) has given guidelines on use of imaging and treatment for headache. These include

1. Don't perform neuroimaging studies in patients with stable headaches that meet criteria for migraine
2. Don't perform computed tomography (CT) imaging for headache when magnetic resonance imaging (MRI) is available, except in emergency settings.
3. Don't recommend surgical deactivation of migraine trigger points outside of a clinical trial.
4. Don't prescribe opioid or butalbital-containing medications as first-line treatment for recurrent headache disorders.
5. Don't recommend prolonged or frequent use of over-the-counter (OTC) pain medications for headache.
   A. **Imaging:** Imaging is necessary in focal neurological findings, altered state of consciousness, Papilledema, vomiting or syncope at the onset of headache, fever (meningitis), neck stiffness (meningitis, SAH), recent headache, progressively worsening headache (SOL), or persisting headache.

1. **CT head (plain and contrast):** It is indicated in emergency scenarios like head trauma, neck trauma,[12] acute thunder clap headache (subarachnoid hemorrhage) since it is widely available, fast, and highly sensitive to blood and fractures. Its sensitivity approaches 100% to blood if performed within 6 hours.[13]

2. **MRI brain (plain and contrast):** It is preferred in non-emergency situations. It is better because of better anatomical delineation and no radiation exposure. It can show subdural fluid collections, enhancement of the pachymeninges, infarcts, engorgement of venous structures, pituitary hyperemia, sagging of the brain, hydrocephalus, space occupying lesions, Chiari malformation which are difficult to appreciate on CT. It is also preferred in pregnancy due to lack of radiation, and better pickup of pituitary apoplexy and cerebral venous sinus thrombosis.

   In cases of headache along with chronic neck pain, a **MRI of cervical spine** should be advised to rule out causes of cervicogenic headache, for example, Chiari malformation.

3. **MRI brain (gradient echo):** If imaging is done after 3–4 days of suspected SAH, MRI-gradient echo is better than CT.[14]

4. **Angiogram (MRA/CTA/DSA):** Cerebral angiogram is needed for aneurysms, AVMs which may mimic migraine. Whenever non-traumatic subarachnoid hemorrhage, or cisternal traumatic hemorrhage is seen on CT/MRI, an angiogram should be advised. In traumatic cases a high index of suspicion is needed.[15] Neck vessel angiogram is advised for carotid dissection, which may occur in association with trauma or fibromuscular dysplasia.[11]

5. **MR Venogram:** It is indicated when there is headache with normal MRI and normal CSF opening pressure (explained below), which is, though not pathognomic, generally seen in cerebral venous sinus thrombosis.

B. Lumbar puncture: Indicated when CT head is normal in above conditions. It helps in diagnosing many conditions including meningitis, neurological disorders (multiple sclerosis, sarcoidosis, etc.), ICP abnormalities (idiopathic intracranial hypertension, normal pressure hydrocephalus, spontaneous intracranial hypotension), and to administer therapeutic agents (antibiotics, chemotherapy, baclofen, contrast for cisternography).[16] It can also be used to diagnose subarachnoid hemorrhage when CT is negative (performed after several days) and suspicion is high.

Before the procedure patient should be told about the possibility of transient post-lumbar puncture headache (32% incidence).[17] Full aseptic precautions (Chlorhexidine in place of Betadine to avoid chemical meningitis) should be taken. It is done in either lying down side turned position (essential when pressure measurement is required), or in sitting position with head and both lower limbs flexed. 2% Lignocaine is injected at L3-4 space at the intersection of midline and intercristal line (also called Tuffier's line—at the level of highest point of iliac spines). The distance from skin to subarachnoid space varies according to age and physique of a person. Needles used can be Quincke, Tuohy, or Whitacre.

CSF pressure is measured by a manometer (Fig. 1.1). Normal values of CSF pressure is 8–15 mm Hg in adults. In infants the range is 0–5 with mean of 2.8 ± 1.4 mm Hg.[18]

Investigations to be advised in a lumbar puncture depends upon the suspicion, but in routine are protein (15–60 mg/dl), sugar (50 to 80 mg/dl—2/3rd of the blood sugar level), cells (0 to 5, with nil RBCs). When infection is suspected or diagnosed appropriate cultures are advised. CSF PCR in the first week can demonstrate herpes simplex virus infection.

Biometrix CF-0112

Fig. 1.1: A manometer for measuring CSF pressure.

C. **Ventricular puncture:** Can be done in infants when the fontanels are open to analyze CSF.
D. **ICP monitoring:** It can serve both diagnostic and therapeutic purposes. It is indicated in mainly traumatic situations. According to brain trauma foundation guidelines, 4th edition, it is advised when: (a) Abnormal CT head with a GCS between 3 and 8, or (b) Normal CT head with any two of the following:

   i. age<40 years
   ii. Unilateral or bilateral motor posturing
   iii. Systolic BP <90 mm Hg.

   In non-traumatic situations it is indicated in Reye's syndrome, subarachnoid hemorrhage with hydrocephalus, posterior fossa masses with hydrocephalus, and idiopathic intracranial hypertension. In pediatric head injuries evidence is not strong but guidelines suggest to monitor if GCS <8.[19]
   Gold standard of ICP monitoring is to place intraventricular catheters which can serve both diagnostic and therapeutic procedures. Other methods include putting intraparenchymal, subdural and epidural bolts. There are non-invasive methods also like tympanic membrane pressure monitoring, optic nerve sheath diameter,[20] venous ophthalmodynamometry,[21] two depth transorbital Doppler, cochlear fluid pressure.
E. **Blood tests:** As per ICHD, for diagnosis of headaches there is no role for blood tests in head or neck trauma, psychiatric disorders, or pain attributed to disorders of cranium, neck, eyes, ears, teeth, mouth or facial structures. Serum tests are useful for diagnosis in GCA, may be helpful in diagnosis of mitochondrial encephalomyopathy, lactic acidosis, and stroke like episodes (MELAS) and cerebral autosomal dominant arteriopathy with subcortical infarcts and leukoencephalopathy (CADASIL) syndromes, and aid in establishing the cause of stroke.[9]
F. **ESR/CRP:** ICHD criteria says either elevated ESR/CRP or biopsy of temporal artery to be diagnostic of GCA. American College of Rheumatology uses ESR to be 50 mm/hr as diagnostic of GCA.
G. **Other serum/urine tests:** Due to rarity of systemic causes of headache the tests should be used judiciously. Also

they may be false positive in inappropriate clinical situations.[22] For systemic and neurologic causes which can cause headaches like

- Anaemia/polycythemia—hemoglobin (Hb), hematocrit
- Hyper/hypothyroidism[23]—T3,T4,TSH may be altered.
- Stroke—glucose, cholesterol, homocysteine, protein C, protein S, factor VIII, antithrombin III, and factor V Leiden mutation.
- Intracranial hypertension secondary to metabolic, toxic or hormonal causes—vitamin A, hormones. Urine can detect anabolic steroids.
- Substance abuse/withdrawal headache—serum and urine levels of alcohol, mercury, lead, cocaine, copper, marijuana.
- Pituitary apoplexy—cortisol, TSH, prolactin, growth hormone
- Infection—serum and urine tests may identify the cause of infection—ICHD
- Systemic lupus erythematosus—raised ANCA, dsDNA apart from low Hb, platelets, and leukocyte count may be found. Urine testing may show proteinuria.
- Genetic testing[22]
  - Familial hemiplegic migraine—CACNA1A, ATP1A2 gene, SCN1A gene
  - CADASIL—notch 3 mutation can be seen on chromosome 19.
  - MELAS—six mtDNA point mutations.

Differential diagnosis: ICHD-3 beta[1] has classified headaches in primary and secondary ones. The details are out of scope of this book, but we try to discuss briefly a few of them.

PRIMARY HEADACHES

1. **Migraine:** In the Global burden of disease 2015, Migraine is listed as one of the eight diseases prevalent in more than 10% of world's population.[24] It may be with or, more commonly, without aura with a ratio of 1:5.[25] Aura may last for 20–30 min which is considerably longer than for a seizure. Typical patient is a young female with unilateral pulsatile headaches in perimenstrual period. Headache, nausea or vomiting may

last for hours or days. Hypersensitivity phenomenon like photophobia, osmophobia, or phonophobia may occur. Precipitating factors may be hunger, lack of sleep, etc. Variants of typical migraine are basilar migraine (associated with brainstem features),[26] ophthalmoplegic or retinal migraine (associated with weakness of extraocular muscles), migraine following head injury, migraine in children, familial hemiplegic migraine, TIAs and stroke with migraine.

2. **Cluster headache:** The typical patient is a young male (M:F ratio 5:1) with severe peri-orbital localization. It is accompanied with vasomotor phenomenon like rhinorrhea, lacrimation, flush and edema of the cheek; generally occurs after 1–2 hours of sleep every night, and this cycle repeats for weeks, only to recur after a year or so.

3. **Tension headache:** It is the most common type of headache. Location is generally bilateral occipitonuchal, dull aching or tightness in character. It is present for prolonged period and gets relieved with sleep. Typical patient is a middle-aged women having anxiety or depression.

SECONDARY HEADACHES

1. **Headache of raised ICP:** Headache in association with visual blurring, papilledema and vomiting is due to raised ICP. It generally occurs in bifrontal region. There is no association with the severity of symptoms and degree of raised ICP.[27] Early symptoms may be irritability, altered behavior, refusal to feed (in children). Headache is often nocturnal or early morning due to two reasons—firstly because of decreased venous return while lying, and vasodilation because of CO_2 retention and increased parasympathetic activity in the night. Pressure over the autonomic structures of brainstem produces the classical features of nausea, vomiting (projectile, i.e. without nausea), hypertension, altered respiration. Presence of hypertension, bradycardia, and slowing of respiration is known as Cushing's reflex. Vomiting relieves the headache due to bringing the patient upright and washing-out of CO_2. ICP can be monitored to diagnose it.

2. **Headaches of brain tumour:** Headache is present in only about 50% of brain tumours.[28,29] Headache occurs due to

mechanical or physiological causes. Mechanical includes mass effect by the tumour over meninges or basal vessels, or producing hydrocephalous due to obstructive or communicating (producing excess of proteins blocking arachnoid villi) causes. They are similar to tension headaches but differ in getting worsened on bending down.[28] Headache is less common in GBM and pituitary tumours, and more common in infratentorial and intraventricular tumours.[29] If one compares tumours of intra-axial or extra-axial, then headache is more common in extra-axial ones due to early contact with meninges. Headaches from brain tumours can occur at any period of the day, and if it causes raised ICP, then it may present with nocturnal, or early-morning headaches.

3. **Subarachnoid hemorrhage (Fig. 1.2):** Typical symptoms described with SAH is sudden severe worst headache of life. Incidence of SAH of all causes of headaches presenting in emergency is <1%.[30] Most common cause of SAH is head trauma. Other causes include aneurysms and arteriovenous malformations. Whenever SAH is encountered in non-traumatic settings, an angiogram (CTA/MRA/DSA) should be advised.

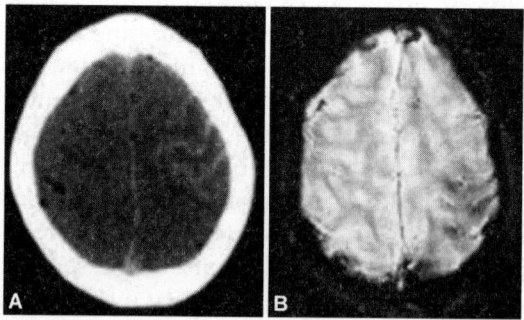

Fig. 1.2A and B: (A) CT head (plain) and (B) MRI brain (GRE) showing subarachnoid hemorrhage

Important Points

- As said by Drs Victor and Adams: "A second examination is the most helpful diagnostic test in a difficult neurologic case."[25]
- Most common prevalent headache : Tension type headache—50–70%

- Prevalence of headache due to raised ICP is <0.01%.
- Intracranial pain sensitive structures: Basal dura, basal arteries (circle of Willis), venous sinuses, cortical veins, intracranial portions of 5th, 9th, 10th, and upper cervical nerves.
- Extracranial pain sensitive structures: Scalp vessels and muscles, orbital contents, mucous membranes of nasal and paranasal spaces, external and middle ear, teeth and gums.
- Mechanism of tension type headache: Muscular due to persistent contraction, e.g. furrowing of brow, head posture, clenching of teeth.
- Dura of anterior and middle cranial fossa are innervated by V1 and V2 nerves (pain referred to forehead or temple—bilateral or ipsilateral).
- Dura of posterior fossa innervated by 9th, 10th, and upper cervical nerves (pain referred to suboccipital or upper cervical region—bilateral or ipsilateral).
- Migraine can be with aura, without aura (more common).
- Headache in <5 years children should be taken seriously.
- Headache is more common in infratentorial tumours than supratentorial tumours.
- Headache due to tumours depends upon tumour type—some say more common in GBM, some say it more common in metastases.
- Patients should be told to maintain a headache diary, as often there "sometimes" headache turn out to be daily headaches changing the management.[11]

REFERENCES

1. No Title. https://www.ichd-3.org/.
2. Schobitz E, Qureshi F, Lewis D. Pediatric headaches in the emergency department. *Curr Pain Headache Rep.* 2006;10(5):391–396.
3. Ravishankar K. The art of history-taking in a headache patient. *Ann Indian Acad Neurol.* 2012;15(Suppl 1):S7-S14. doi:10.4103/0972-2327.99989.
4. Wolff V, Ducros A. Reversible Cere bral Vasoconstriction Syndrome Without Typical Thunderclap Headache. *Headache.* 2016;56(4):674–687. doi:10.1111/head.12794.

5. Mainardi F, Maggioni F, Zanchin G. The Case of the Woman Who Did Never Dare to Fly: Headache Attributed to Imbalance Between Intrasinusal and External Air Pressure. *Headache*. 2016;56(2):389–391. doi:10.1111/head.12766.

6. Folstein MF, Folstein SE, McHugh PR. "Mini-mental state". A practical method for grading the cognitive state of patients for the clinician. *J Psychiatr Res*. 1975;12(3):189–198.

7. Ganguli M, Ratcliff G, Chandra V, et al. A Hindi version of the MMSE: the development of a cognitive screening instrument for a largely illiterate rural elderly population in India. *Int J Geriatr Psychiatry*. 1995;10(5):367–377.

8. Walker H, Hall W, Hurst J, eds. *Clinical Methods: The History, Physical, and Laboratory Examinations*. 3rd ed. Butterworth Publishers, a division of Reed Publishing; 1990.

9. Dubois B, Slachevsky A, Litvan I, Pillon B. The FAB: a Frontal Assessment Battery at bedside. *Neurology*. 2000;55(11):1621–1626.

10. Dodick D. Diagnosing Headache: Clinical Clues and Clinical Rules. *Adv Std Med*. 2003;3(2):87–92.

11. Nye BL, Ward TN. Clinic and Emergency Room Evaluation and Testing of Headache. *Headache*. 2015;55(9):1301–1308. doi:10.1111/head.12648.

12. Bajaj J, Mittal RS, Sharma A. Epidemiology of Spinal Injuries: An Experience in tertiary and Regional Referral Hospital of Northwest India. *J Spinal Surg*. March 2014.

13. Perry JJ, Stiell IG, Sivilotti MLA, et al. Sensitivity of computed tomography performed within six hours of onset of headache for diagnosis of subarachnoid haemorrhage: prospective cohort study. *BMJ*. 2011;343:d4277.

14. Mitchell P, Wilkinson ID, Hoggard N, et al. Detection of subarachnoid haemorrhage with magnetic resonance imaging. *J Neurol Neurosurg Psychiatry*. 2001;70(2):205–211.

15. Bajaj J, Agrawal M, Sinha VD. Blunt orbital injury causing traumatic intracranial aneurysm in a child. 2016. doi:10.4103/0976-3147.165436.

16. Doherty CM, Forbes RB. Diagnostic Lumbar Puncture. *Ulster Med J*. 2014;83(2):93–102.

17. Armon C, Evans RW. Addendum to assessment: Prevention of post-lumbar puncture headaches: report of the Therapeutics and Technology Assessment Subcommittee of the American Academy of Neurology. *Neurology*. 2005;65(4): 510–512. doi:10.1212/01.wnl.0000173034.96211.1b.

18. Kaiser AM, Whitelaw AG. Normal cerebrospinal fluid pressure in the newborn. *Neuropediatrics*. 1986;17(2):100–102. doi:10.1055/s-2008-1052509.

19. Adelson PD, Bratton SL, Carney NA, et al. Guidelines for the acute medical management of severe traumatic brain injury in infants, children, and adolescents. Chapter 5. Indications for intracranial pressure monitoring in pediatric patients with severe traumatic brain injury. *Pediatr Crit Care Med*. 2003;4(3 Suppl):S19-24.

20. Cennamo G, Gangemi M, Stella L. The correlation between endocranial pressure and optic nerve diameter: an ultrasonographic study. In: *Ophthalmic echography*. Springer; 1987:603–606.

21. Baurmann M. Über die Entstehung und klinische Bedeutung des Netzhautvenenpulses. *Dtsch Ophthalmol Ges*. 1925;45:53–59.

22. Loder E, Cardona L. Evaluation for secondary causes of headache: the role of blood and urine testing. *Headache*. 2011;51(2):338–345. doi:10. 1111/j.1526–4610.2010.01840.x.

23. Larner AJ. Thyroid dysfunction and headache. *J Headache Pain*. 2006;7(1):51–52. doi:10.1007/s10194-006-0263–9.

24. Global, regional, and national incidence, prevalence, and years lived with disability for 310 diseases and injuries, 1990-2015: a systematic analysis for the Global Burden of Disease Study 2015. *Lancet (London, England)*. 2016;388 (10053): 1545-1602. doi:10.1016/S0140-6736(16)31678-6.

25. Ropper A, Samuels M, Klein J. *Adam and Victor's Principles of Neurology*. 10th ed. New York: McGraw-Hill; 2014.

26. Bickerstaff E. Basilar artery migraine. *Lancet*. 1961;277(7167):15–17.

27. Dunn LT. Raised intracranial pressure. *J Neurol Neurosurg Psychiatry*. 2002;73 Suppl 1:i23–7.

28. Forsyth PA, Posner JB. Headaches in patients with brain tumors: a study of 111 patients. *Neurology*. 1993;43(9):1678–1683.

29. Pfund Z, Szapary L, Jaszberenyi O, Nagy F, Czopf J. Headache in intracranial tumors. *Cephalalgia*. 1999;19(9):787–90; discussion 765. doi:10.1046/j.1468-2982.1999.1909787.x.

30. Morgenstern LB, Huber JC, Luna-Gonzales H, et al. Headache in the emergency department. *Headache*. 2001;41(6):537–541.

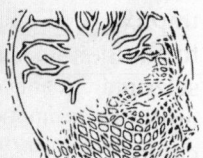

Seizures

```
┌─────────────────────┐
│      Seizures       │
└─────────────────────┘
```

History

- Age of onset
- Type and semiology
- Loss of consciousness
- Duration of a single episode
- Number of episodes
- Aura
- Ictal phase
- Post-ictal phase
- Diurnal variation
- Trigger factors
- Associated lobar features
- Drugs
- Trauma
- Perinatal history
- Febrile convulsions
- Family history
- CNS infections

Examination

- General examination
- Skin examination
- Neurologic examination
- Ophthalmic examination
- Cardiovascular examination
- Neuropsychological examination

Seizure literally means the action of capturing someone or something using force. It can be of focal onset, generalised onset or unknown onset. According to International League Against Epilepsy (ILAE), **Epilepsy** is defined as two or more unprovoked seizures more than 24 hours apart; or one unprovoked seizure and a probability of further seizures similar to the general recurrence risk (60%) after two unprovoked seizures, occurring over the next 10 years; or diagnosis of an epilepsy syndrome. Through this chapter we will try to clinically localise the seizure onset zone, and learn about its appropriate investigations.

History

- **Age of onset:** Neonatal seizures may occur because of perinatal insult, metabolic disorder, congenital malformation, intracranial haemorrhage. In adults and teenagers, common causes are idiopathic, mesial temporal sclerosis, cortical dysplasias, trauma, tumours.
- **Type:** Generalized (associated with loss of consciousness) tonic clonic, tonic (common in frontal, less in temporal), absence, atonic, myoclonic, partial (consciousness preserved), complex partial (consciousness impaired, may lead to secondary generalization) (Table 2.1).

Table 2.1: Semiology and types of seizures, clinical manifestation, localizing and lateralizing value

Type	Clinical manifestation	Localization	Lateralization
Tonic	Tonic spasms	Temporal, frontal, supplementary sensorimotor area	Temporal (only unilateral), frontal (bilateral).[1] Tonic seizures from supplementary sensorimotor area are bilateral[2]
	Tonic elevation of arms, flexion of elbows	Secondary motor area	Contralateral hemisphere
Autonomic	Ictal tachycardia piloerection	Temporal	Poor
Clonic	Clonic movements	Frontal, temporal	If unilateral, then contralateral hemisphere. High predictive value

Contd.

Table 2.1: Semiology and types of seizures, clinical manifestation, localizing and lateralizing value (Contd.)

Type	Clinical manifestation	Localization	Lateralization
End of seizure paradoxical clonus	Seizure activity persists longer on the side of epileptogenic focus because the ipsilateral cortex exhausts early	Frontal, temporal	High predictive value
Tonic clonic	Tonic clonic movements	Generalized	Poor
Versive	Turning of face and eyes in one direction	Contralateral frontal lobe, less likely temporal lobe	High
Hypermotor	Pedalling, running, pelvic thrusting	Orbital or mesial frontal regions. Less likely for insular and temporal regions	Less
Automatic	Distal segments of hands, feets, mouth, tongue	Temporal, frontal	Unilateral involvement points to non-dominant mesial temporal lobe epilepsies[3]
Dialeptic seizures	Amnesia of the event of seizure	Poor	Poor
Gelastic seizures	Excessive laughing	Hypothalamic hamartoma	Poor
Atonic	Loss of tone	Generalized (Lennaux Gestaut syndrome), frontal and temporal[4]	Poor
Astatic seizure	Epileptic falls (Myoclonic	Unknown	Unknown

Contd.

Table 2.1: Semiology and types of seizures, clinical manifestation, localizing and lateralizing value (Contd.)

Type	Clinical manifestation	Localization	Lateralization
	seizure foll-owed by atonic seizure)		
Akinetic activity	Absence of motor	Mesial frontal and inferior frontal gyri	Poor
Aphasic	Aphasia despite preserved awareness and memory	Dominant lobe	High

- Todd's paralysis is a highly lateralizing sign. The motor signs and paralysis occur contralateral to the epileptogenic hemisphere.

- Unilateral eye blinking has a high lateralizing value. It occurs ipsilateral to the epileptogenic focus. No localizing value.

- **Ictal nystagmus:** Fast component opposite the seizure focus. Origin is temporal or temporo-occipital junction.

- **Loss of consciousness:** Consciousness is lost in generalised seizures, and focal seizures with secondary generalisation. Preservation of consciousness during bilateral seizures localizes the area to be supplementary motor cortex.

- **Duration of a single episode:** Seizures of mesial temporal lobe are of longer duration, while frontal lobe seizures are of shorter one.[5] Myoclonic seizures last 200 msec, most often seen in generalized epilepsies.[6]

- **Number of episodes:** Recurrent episodes (more than two 24 hours apart) dignify it to be a epilepsy.

- **Other semiology:** Excessive blinking (occipital lobe), staring gaze (absence seizure). Similar semiology of a lesion identifies it to be of single pathology, apart from helping to localising it.

- **Aura:** Common in partial seizures (Table 2.2).

Table 2.2: Types of auras, clinical symptoms, localizing and lateralizing value

Type of Aura	Clinical symptoms	Localizing value	Lateralizing value
Sensory	Tingling/numbness/insect crawling	Somatosensory cortex	Contralateral somatosensory cortex if unilateral symptoms, and secondary somatosensory cortex if bilateral symptoms
Olfactory	Hallucinations of unpleasant smell	Mesial temporal lobe	None
Gustatory	Unpleasant taste	Insula[7]	None
Autonomic	Palpitations, sweating, goose bumps	Insular cortex[3]	None
Abdominal	Nausea, tenseness, knot, squeezing	Temporal (mostly), frontal, insular	None
Psychic	Autoscopy, fear, Deja vu, Jamais vu, visual hallucinations	Temporal Parieto-temporal, temporo-occipital	None
Visual	Flashes of light	Broadmann's area 17, 18	
Auditory	Auditory hallucinations	Heschell's gyrus	

- **Ictal phase:** Patient may have a variety of semiology as mentioned in Table 2.1. In generalised seizures there will be loss of consciousness. Patient may have urination or defecation, and can suffer trauma. Focal seizures will show preserved consciousness.
- **Post-ictal phase:** Patient may complain of headache, confusion, disorientation after generalised seizures. No confusion/disorientation is seen after focal seizures.
- **Diurnal variation:** Early morning (myoclonic), nocturnal sometimes (frontal), anytime (temporal)
- **Triggering factors:** Non-compliance of drug, sleep deprivation, alcohol consumption, stress, fever, exercise.

- **Associated lobar features:** For localization
- **Drugs:** One should enquire about the drugs that are being taken by the patient. Certain drugs which precipitates or increases the propensity of seizures are: Fluoroquinolones (ofloxacin, levofloxacin, gatifloxacin), imepenem, alcohol, heroin, etc.
- **Trauma:** Head injury may lead to foci for epilepsy.
- **Perinatal history:** Birth hypoxia is a frequent cause of epilepsy.
- **Febrile convulsions:** These if present in childhood may lead to epilepsy in adulthood.
- **Family history:** Any incident of seizures or neurological illness should be enquired for.
- **CNS infections:** Meningitis, encephalitis history should be asked.

EXAMINATION

General examination: To look for systemic causes of seizures (hypoglycaemia, electrolyte abnormalities).

Skin examination: Often neurocutaneous syndromes are associated, for example, tuberous sclerosis (adenoma sebaceum, shagreen patches, depigmented nevi, facial angiofibromas), Sturge-Weber syndrome (facial nevus flammeus), neurofibromatosis. Dysmorphic syndromes like Proteus syndrome (skin overgrowth, tumours over half of body, abnormal bone development) may be associated. Abnormal pigmentation like incontinentia pigmenti may be associated. Facial asymmetry may be associated with temporal lobe epilepsy.[1]

Neurologic: Complete neurologic examination including MMSE, cranial nerves, sensorimotor examination should be done. One should note down after what interval of the seizure, the examination was done.

- Examined immediately after the seizure then focal signs such as Todd's paresis, transient aphasias should be looked for.
- Examined after a considerable time, then one should look for permanent dysfunctions and focal brain lesions. There may be presence of language disturbances, hemiparesis, cognitive decline.

- Tests of laterality should be performed: Luria's test (as explained in Chapter 1), because often the handedness shifts to opposite hemisphere in epilepsy.

Ophthalmic examination: Papilledema and field cuts should be examined which may be seen in raised ICP, and focal brain lesions respectively. Oculomotor ophthalmoplegia may be seen in brainstem, cavernous sinus lesions.

Cardiovascular examination: Causes of syncopal attacks like arrhythmia, murmurs should be sought for.

Neuropsychological examination: Latent anxiety, depression, psychosis should be searched for. Also, baseline cognitive and IQ assessment is essential before any epilepsy surgery.

INVESTIGATIONS

A. **Non-invasive investigations:** These are sufficient in more than 70% of patients.
   1. **Electro-encephalogram (EEG):** This is a must in every seizure/epilepsy case. This also differentiates seizures from pseudoseizures.
   2. **Video EEG:** If a patient is pharmaco-resistant then for his presurgical evaluation VEEG is must. It delineates the semiology better, and can differentiate between seizures and pseudoseizures.
   3. **MRI brain (epilepsy protocol):** This is also mandatory for presurgical evaluation of a patient. Whenever possible, 3-Tesla MRI is preferred. In the oblique coronal orientation, 3 mm slices perpendicular to the plane of hippocampus, with T1 , T2 and Coronal T2 FLAIR, 3D FLAIR, Coronal T2 GRE sequences of entire brain are done. This can delineate hippocampal sclerosis (loss of hippocampal volume, temporal horn dilation, hyperintensity on T2 FLAIR, atrophy of hippocampal efferents, i.e parahippocampal gyrus and fornix), focal cortical dysplasias (loss of grey white differentiation in type I, and transmantle sign in type II), Rasmussen's encephalitis, hemimegalencephaly, etc. Classical findings of hippocampal sclerosis are loss, if a tumour or mass is suspected, then only gadolinium contrast is needed.

4. **Positron emission tomography (PET):** It is not needed in all cases, and is needed when clinical semiology, MRI, and VEEG are non-concordant with each other. It may show hypo-metabolism in the area which is producing seizures.

5. **Ictal single photon emission computed tomography (SPECT):** Ictal SPECT shows hypermetabolism in the epileptogenic zone. It has a limitation that the tracer has to be injected within a short time (maximum 45 seconds after start of seizure).

6. **Functional MRI (fMRI):** It is done for speech, motor, and memory function localisation in the brain, which will help in preservation of these areas during surgery. This has replaced WADA testing in most centres across the world.

7. **Magnetic resonance spectroscopy (MRS):** It can be used to screen for metabolic dearrangements like mitochondrial diseases, creatine deficiencies, etc. It can be used to differentiate between a tumour and focal cortical dysplasia (FCD)—choline/creatine ratio is elevated in tumour, while it is normal in FCD.

8. **Magnetoencephalogram (MEG):** It gives a source localisation of EEG dipoles. It can also give functional limitations. Firstly, it is mostly interictal except when patient is having many seizures or has coincidental seizure at the time of acquisition, and secondly, its cost is high.

B. **Invasive investigations:** These are phase II investigations. They are needed when the built hypothesis of localisation of EZ has insufficient evidence. They are indicated in the following scenarios:

a. Non-concordant investigations

b. Two pathologies on MRI

c. MRI negative epilepsy.

Types of invasive investigations are:

a. **Subdural grid implantation:** In this, a grid of appropriate size (4 × 1, 5 × 4, etc.) is placed after making a craniotomy, and then wound is closed. Patient is monitored and seizures are recorded. It is better than surface EEG since the interference of dura, bone and scalp gets abolished. It has limitations of difficulty in recording the subcortical and deep structures.

b. **Stereotactic depth EEG placement:** In this, electrodes are implanted stereotactically. It has made recording simpler since there is no need of craniotomy, less complications, and can record depth structures.

DIFFERENTIAL DIAGNOSIS

1. **Metabolic disturbances:** These are the most common causes. Hypoglycemia, hypocalcemia, sodium imbalances, magnesium, renal profile dearrangements, etc. are common ones.

2. **Infections:** Neurocysticercosis, meningitis, encephalitis can precipitate seizures and can lead to epilepsy.

3. **Tumours:** Supratentorial and intra-axial tumours are more commonly associated with epilepsy. In them frontal and temporal tumours have higher incidence. Oligodendroglioma, Ganglioglioma, Dysembryoplastic neuroepithelial tumours commonly present with seizures.

4. **Trauma:** Head injury leading to parenchymal contusions, subarachnoid hemorrhage have preponderance than other types of head injury to produce seizures.

5. **Developmental malformations:** Focal cortical dysplasias are a common cause of drug resistant epilepsy in children. Schizencephaly, porencephaly, hemimegalencephaly are other developmental disorders associated with epilepsy.

6. **Mesial temporal sclerosis:** This is a common cause of drug resistant epilepsy in adults. This has a good seizure freedom outcome rates after surgery.

7. **High fever:** Febrile convulsions can occur with high grade temperature (more than 100°C) in children between 6 months and 5 years.

8. **Drugs:** Alcohol, cocaine, heroine, fluroquinolones, imepenem, amphetamines, street drugs, etc. can cause seizures.

9. **Stroke:** This may lead to seizures

10. **Preclampsia:** This is associated with seizures.

11. **Bites and stings:** Snake bite, scorpion stings may cause seizures.

REFERENCES

1. Werhahn KJ, Noachtar S, Arnold S, et al. Tonic seizures: their significance for lateralization and frequency in different focal epileptic syndromes. *Epilepsia.* Sep 2000;41(9):1153–1161.
2. Ohara S, Ikeda A, Kunieda T, et al. Propagation of tonic posturing in supplementary motor area (SMA) seizures. *Epilepsy research.* 2004;62(2):179–187.
3. Tufenkjian K, Lüders HO. Seizure semiology: its value and limitations in localizing the epileptogenic zone. *Journal of Clinical Neurology.* 2012;8(4):243–250.
4. So NK. Atonic phenomena and partial seizures. A reappraisal. *Advances in neurology.* 1995;67:29–39.
5. Wyllie E, Luders H, Morris HH, Lesser RP, Dinner DS. The lateralizing significance of versive head and eye movements during epileptic seizures. *Neurology.* 1986;36(5):606–611.
6. Vendrame M, Zarowski M, Alexopoulos AV, Wyllie E, Kothare SV, Loddenkemper T. Localization of pediatric seizure semiology. *Clinical neurophysiology : official journal of the International Federation of Clinical Neurophysiology.* Oct 2011;122(10):1924–1928.
7. Stephani C, Vaca GF-B, Maciunas R, Koubeissi M, Lüders H. Functional neuroanatomy of the insular lobe. *Brain Structure and Function.* 2011;216(2):137–149.

3

Speech

```
                    ┌──────────────┐
                    │    Speech    │
                    └──────────────┘
              ┌───────────┴───────────┐
```

History

- Onset, duration and progression
- Native language, literacy
- Focal neurological deficits
- Oral problems

Examination

- MMSE/higher mental functions
- Mutism
- Articulation
- Fluency
- Comprehension
- Reading
- Writing
- Naming
- Repetition
- Head and neck examination
- Exclude oral problems
- Check laryngeal functions
- Cranial nerves
- Focal neurological deficits
- Neuropsychological examination

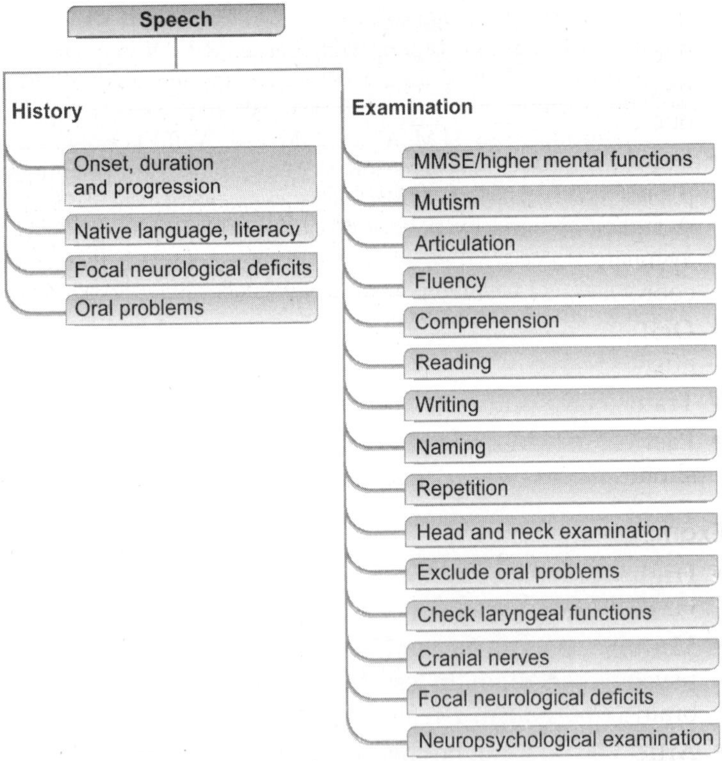

For speech we have to take history and examine for lobar (Wernicke's, Broca's, conduction, etc.), cerebellar, lower cranial nerves, pharynx, tongue, and lips involvement.

We should divide speech disorders into four types: Psychogenic (stammering, lisping, lalling), dysarthria (defect in articulation), aphasia/dysphasia (defect in language; dominant hemisphere disorders), aphonia/dysphonia (laryngeal disorders).

History

- **Onset, duration and progression:** Congenital onset causes are like stammering, lisping, lalling (baby speech), etc.
- **Native language, handedness, literacy:** Native language should be enquired from the patient. It has been observed and derivated from Ribot law of retained distant memory that, after aphasia, individuals with proficiency in more than one language improved earlier in their native language. Pitres law states that the language used more often is recovered first. To determine side of cereberal dominance one can enquire of handedness for writing, sewing, etc.
- **Focal neurological deficits:** Enquire about cortical symptoms and signs like hemiparesis, hemianopia, field defects. A right faciobrachial weakness coordinates with Broca's aphasia. Similarly, right hemianopia or quadrantanopia coordinates with Wernicke's aphasia.
- **Oral problems:** It should be asked, as any oral problem can affect speech.
- **Trauma and surgery:** History of these should be enquired.
- **Past history:** Medical disorders like hypertension, diabetes should be enquired, particularly in sudden onset ones.

Examination

- **Higher mental functions:** Examine MMSE, and detailed HMF if there is doubt of cortical involvement.
- **Mutism:** If a normal patient is mute, it is suggestive of psychogenic origin. In lesions of cerebrum, cerebellum, and brainstem, mutism can also occur.
- **Articulation:** It can be checked by listening to the patient or making him reading a book loud. Distinguish the types of dysarthria. First make him speak labials (papa, mama) for finding out facial weakness (peripheral facial palsy, myopathies of facial muscles). Then make him speak linguals (tata, dada) for anterior tongue (hypoglossal) weakness, and then make him speak velars (kaka) for palatal or posterior tongue weakness. If the speech is hypernasal,

then it is because of palatal weakness (Xth nerve lesion). It is checked by keeping a metal blade in front of nose and asking the patient to speak ring, sing which will produce increased fogging. Scanning speech may be seen in cerebellar lesions.

- **Fluency:** To start, say to the patient: 'Tell me something about you.' This will get them talking and enable you to screen for any abnormalities. Fluency, praxis, prosody, and language formulation can be assessed by engaging into a talk during history taking.
- **Comprehension:** Ask the patient to perform 3-step commands. Impaired comprehension is seen in Wernicke's aphasia, global aphasia. Impaired comprehension but normal formulated speech and ability to read suggests rare syndrome of pure word deafness.
- **Reading:** When conversation is normal but there is impaired readability, then it is suggestive of pure word blindness.
- **Writing:** It is impaired in all speech disorders except in pure word blindness.
- **Naming:** Ask the patient to name and describe objects and what they do (nominal aphasia/expressive dysphasia). Loss of naming ability is found in almost all type of aphasias.
- **Repetition:** Ask the patient to repeat the words: 'Ramgopalacharya'—listen for dysarthria/bulbar involvement (LMN = nasal speech + tongue fasciculation + wasting), pseudobulbar (UMN = hot potato speech + increased jaw jerk + spastic). Repetition is impaired in all forms of aphasias except in transcortical aphasias (because of intactness of perisylvian area) and anomic aphasia.
- **Head and neck examination:** It is important as disorders of thyroid, and other neck structures can cause speech disorders by compression of the nerves.
- **Exclude oral problems:** This should be excluded always.
- **Check for laryngeal disorders (dysphonia):** Ask the patient to cough. Normal cough suggests normal vocal cord function. Mirror examination, direct laryngoscopy, videoendoscopy, and videostroboscopy[1] can be done.
- **Cranial nerves:** Examine VIIth, IXth, Xth, XIth, and XIIth nerves to localise the lesion.

- **Focal neurological deficits:** Examine these, as these may be associated signs.
- **Neuropsychological examination:** It should be done if there is doubt for hysteria.

Investigations

- **MRI Brain:** It is done when an intracranial pathology is suspected. DWI imaging can be added when stroke is suspected, and contrast can be added when a space occupying lesion is suspected.
- **Laryngeal EMG (LEMG):** It measures the nerve inputs to laryngeal muscles.
- **Voice lab functional testing:** It is done by sound pathologist, that determines different speech parameters through acoustic analysis.
- **Test for reflux:** Prolonged reflux can cause voice disorders. To rule it out tests like double pH monitoring can be advised.

Types of Dysphasias/Aphasia[2-4]

- **Broca's aphasia:** Speech is slow, nonfluent, short phrased, agrammatical, but transmits ideas. Comprehension is normal. Repetition is impaired. Right arm and facial weakness is often associated. Localises to Broca's area damage in inferior frontal gyrus.
- **Wernicke's aphasia:** Comprehension is impaired. Speech is fluent, voluminous, and lacks meaning. Repetition is none. Right hemianopia or quadrantanopia often accompanies. There is no paresis. Localises to Wernicke's area damage in speech dominant superior temporal gyrus.
- **Conduction aphasia:** There is normal comprehension and fluency, repetition is impaired, and no associated signs. It localises to arcuate fasciculus damage in supramarginal gyrus or insula.
- **Global:** Speech is nonfluent, comprehension is impaired, and repetition is impaired. Hemiplegia is a usual sign. It localises to large perisylvian area damage or frontal/temporal damage.
- **Transcortical motor:** Speech is nonfluent, comprehension is preserved, and repetition is preserved. Variable associated

signs. It localises anterior or superior to Broca's area damage. Repetition is preserved because of sparing of arcuate fasciculus.

- **Transcortical sensory:** In this, fluency is preserved, comprehension is defective, and repetition is preserved, with variable associated signs. It localises to surrounding areas of Wernicke's area.
- **Transcortical mixed:** Both Broca's and Wernicke's area are damaged, thus impairing both fluency and comprehension, but repetition is spared.
- **Pure word deafness:** Speech is mildly paraphasic or normal. Comprehension is impaired. Repetition is impaired. Associated signs are absent or there may be mild quadrantanopia. It localises to bilateral or left middle of superior temporal gyrus.
- **Pure word blindness:** Alexia without agraphia. Patient is unable to read his own writing. Fluency, comprehension and repetition are normal. Associated signs may be right hemianopia. It localises to left calcarine cortex, or callosum or angular gyrus.
- **Pure word mutism (aphaemia):** Patient is mute, but able to write. There is normal comprehension, absent repetition, with no associated signs. It localises to region around Broca's area.
- **Anomic aphasia:** Isolated naming and word finding difficulty. Comprehension is normal, repetition is normal. Associated signs are generally absent; localises to left deep temporal lobe.
- **Foreign accent syndrome:** It occurs with a stroke with a mild form of Broca's aphasia. Accent may be misinterpreted by the listener who is known to that language, e.g. French, German, etc. This syndrome can be mistaken as hysteria or psychosis.

Types of Dysarthria

- **Bulbar:** There is indistinct and nasal quality of the voice. The site of involvement is brainstem causing LMN type of paresis
- **Pseudobulbar:** There is indistinct, breathy and mumbling quality of voice. There is intermingling of words (spastic

speech). The site of involvement is pyramidal tracts causing UMN type of paresis.

- Scanning speech: There is slurring, and sing-song quality of speech. The site of involvement is cerebellum.
- Myopathic: The site of involvement is muscles of speech.
- Parkinsonian speech: There is stammering speech. Speech is low volume and monotonous. The site of involvement is basal ganglia.

REFERENCES

1. Zacharias SR, Brehm SB, Weinrich B, Kelchner L, Tabangin M, de Alarcon A. Feasibility of Clinical Endoscopy and Stroboscopy in Children With Bilateral Vocal Fold Lesions. *American journal of speech-language pathology.* Nov 01, 2016;25(4):598–604.

2. *Adam and Victor's Principles of Neurology.* 10th ed: McGraw-Hill; 2014.

3. Campbell WW. *Dejong's the neurologic examination* 7th ed: Lippincott Williams and Wilkins 2013.

4. Gami N. *Bedside approach to Clinical Neurology.* 2nd ed: Jaypee.

4

Vomiting

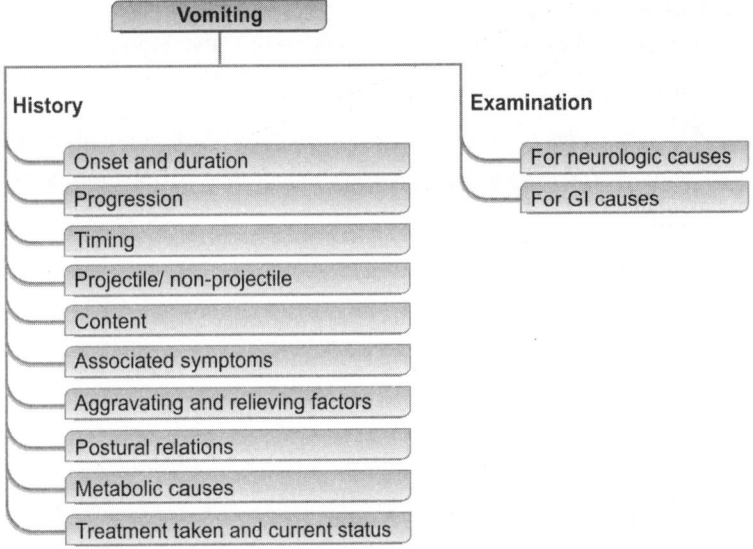

Vomiting	
History	**Examination**
Onset and duration	For neurologic causes
Progression	For GI causes
Timing	
Projectile/ non-projectile	
Content	
Associated symptoms	
Aggravating and relieving factors	
Postural relations	
Metabolic causes	
Treatment taken and current status	

Vomiting: Expulsion of gastrointestinal (GI) contents orally

Regurgitation: Effortless passage of gastric contents into the mouth.

Rumination: Repeated regurgitation of gastric contents which may be rechewed and reswallowed.

CNS control of vomiting: Vomiting centre is present in the medulla. There are four main areas of input to it: GI tract, area postrema, vestibular apparatus, and cerebral cortex. Area postrema senses chemical stimuli, and is influenced by dopamine and serotonin. Vestibular apparatus senses motion and body position, and vomiting is mediated by histamine and acetylcholine. Discrete pathways of cereberal cortex leading to vomiting are not well understood, but they can be strong in stimulating or suppressing vomiting.

History

- **Onset and duration:** Vomiting is generally acute in onset. Chronic recurrent vomiting may be a part of migraine, or cyclic vomiting syndrome (CVS) characterised by symptom free weeks between days of nausea and vomiting (a type of abdominal migraine).

- **Progression:** Progressive vomiting is a bad sign, and is indicative of increasing severity of the causative factor.

- **Timing:** Vomiting due to raised intracranial pressure (ICP) generally occurs in early morning hours because of ICP elevation occurring at that time, while GI vomiting generally occurs after eating (within one hour in pyloric stenosis, later in intestinal obstruction).

- **Projectile/non-projectile:** Projectile vomiting means a type of vomiting in which there is no nausea, and vomitus exits through great force. It occurs in raised ICP and pyloric stenosis.

- **Content:** Sometimes, undigested food comes back into the mouth, and is known as regurgitation, not vomiting. Colour of the vomitus also indicates contents and point towards the possible cause, for example, red coloured vomitus suggests oesophageal bleeding, and green coloured vomitus suggests intestinal obstruction. It lacks the force of vomiting. Vomiting of feculent material suggests distal intestinal obstruction or cologastric fistulae. Contents in gastroparesis is food residue eaten hours to days previously.

- **Associated symptoms:** Headache preceding the vomiting and getting relieved with it is most likely because of raised ICP. Similarly, decreased vision, neurological deficits may suggest raised ICP, tumour or CVA. Epigastric burning or cramping pain in abdomen at the time of vomiting is because of GI cause. Fever, rash suggests meningitis. Decreased hearing, vertigo, tinnitus along with vomiting may be seen in vestibular schwannoma cases or labyrinthine disease.

- **Aggravating and relieving factors:** Motion may precipitate vomiting (because of vestibular imbalance). Drugs may precipitate vomiting: Alcohol, SSRIs, phenytoin, opioids (tramadol), chemotherapeutics, emetics (salt, mustard, copper sulphate, hydrogen peroxide)
- **Postural relations:** Postural changes may lead to vomiting in vestibular disorders. Nausea with light-headedness may occur on getting upright from supine position: It is seen in postural orthostatic tachycardia syndrome (POTS).
- **Metabolic diseases:** One should enquire about diseases like uraemia, hypercalcemia, adrenal insufficiency, hyperglycemia, or hypoglycemia. These may cause repeated vomiting.
- **Treatment taken and current status:** Previous abdominal surgical history may be present. Treatment history is to be enquired to rule out drug-induced vomiting.

Examination

- **Higher mental functions:** Examine for higher mental functions if cortical, subcortical, or raised ICP is suspected.
- **Cranial nerves:** Examine all cranial nerves. Specially see papilledema, visual field abnormalities.
- **Sensorimotor examination:** See for focal neurologic deficits.
- **For GI causes:** Examine the abdomen.

INVESTIGATIONS

- **MRI brain:** It is done whenever an intracranial cause is suspected. It can reveal any space occupying lesion, hydrocephalous, etc.
- **Lumbar puncture:** It can be ordered when meningitis is suspected.
- **USG abdomen:** It is ordered when an intra-abdominal cause is suspected. Free fluid, excessive gas, etc. can be seen.
- **X-ray abdomen** (standing view): It is also ordered when an intra-abdominal cause is suspected. Findings for gut perforation, obstruction are characteristic.
- **Blood investigations:** Like sugar, electrolytes in appropriate situations.

DIFFERENTIAL DIAGNOSIS[1,2]

- *Central nervous system causes*: Raised ICP, migraine, CVSs, meningitis, irritation of area postrema, seizure disorders, emotional responses. Raised ICP leads to vomiting possibly because of transmitted pressure to the area postrema.
- Gastrointestinal cause[3]: Gastroenteritis, gastric outlet obstruction, bowel perforation, intestinal obstruction, pancreatitis, etc.
- *Drugs*: Alcohol, SSRIs, phenytoin, opioids (tramadol), chemotherapeutics, emetics (salt, mustard, copper sulphate, hydrogen peroxide).
- *Labrynthine disorders*: Motion sickness, labrynthitis, Meniere's disease, tumours.
- *Cyclical vomiting syndrome (CVS)*: It is a disorder with repetitive attacks of severe nausea, vomiting, and physical exhaustion with no apparent cause. It can last from a few hours to days.
- *Metabolic disturbances*: Uraemia, hypercalcemia, adrenal insufficiency, hyperglycemia, hypoglycemia.
- Exposure to radiation
- Paraneoplastic syndromes
- *Pregnancy*: 70% have vomiting in 1st trimester.
- *Psychiatric diseases*: Psychogenic vomiting, anxiety, depression, pain, anorexia nervosa, bulimia nervosa

REFERENCES

1. https://en.wikipedia.org/wiki/Vomiting
2. http://bestpractice.bmj.com/topics/en-gb/631
3. Harrison's Principle of Internal Medicine 18th edition.

CHAPTER

5

Anosmia

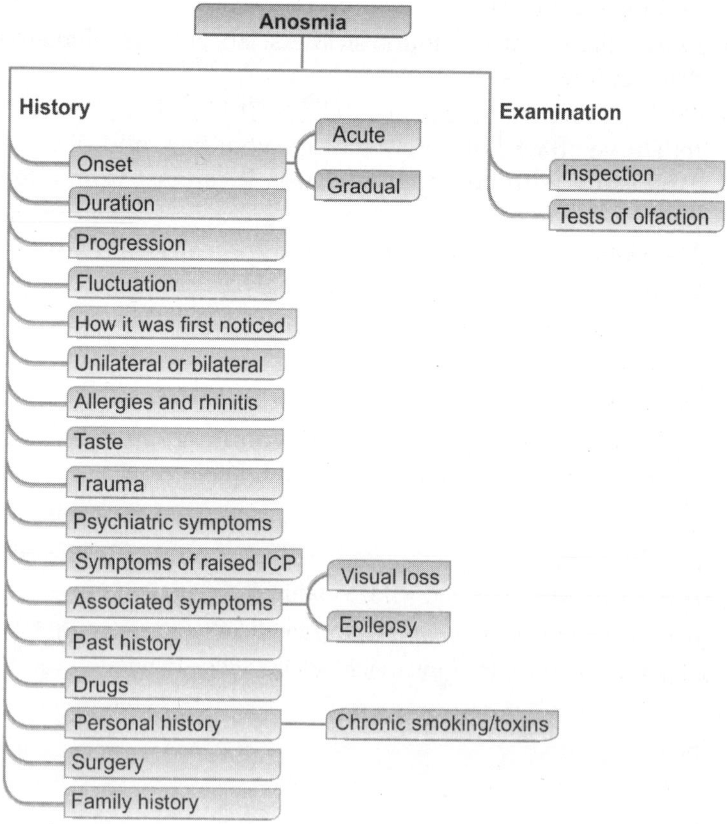

```
                        ┌─────────────────┐
                        │     Anosmia     │
                        └─────────────────┘
```

History

- Onset ─── Acute / Gradual
- Duration
- Progression
- Fluctuation
- How it was first noticed
- Unilateral or bilateral
- Allergies and rhinitis
- Taste
- Trauma
- Psychiatric symptoms
- Symptoms of raised ICP
- Associated symptoms ─── Visual loss / Epilepsy
- Past history
- Drugs
- Personal history ─── Chronic smoking/toxins
- Surgery
- Family history

Examination

- Inspection
- Tests of olfaction

Definition

Anosmia: Complete loss of smell

Hyposmia: Partial loss of smell

Parosmia: Distorted sense of smell (present in viral infection).[1]

Phantosmia: Smelling things when it is absent (present in sinus infection).

There are seven primary odours (ether, peppermint, pungent, putrid, floral, musky, camphor) and thousands of secondary odours.

History

One should take history in a way to find out whether the cause is conductive or sensorineural.

- **Onset—acute or gradual** should be asked. Acute onset can occur in URTIs, head injuries (because of fractured cribriform plate), toxins, and surgery. Gradual ones may occur in ageing, neurodegenerative or autoimmune conditions, and neoplastic lesions like olfactory groove meningioma. Anosmia may be since birth, and is then known as congenital anosmia. Patients with genetic abnormalities give no recollection of smell.

- **Duration:** Short duration are more likely because of conductive problem like rhinitis, while longer duration are more likely because of neural lesion like neoplasm.

- **Progression:** Conductive lesions are generally non-progressive, while neural lesions are more likely to be progressive.

- **Fluctuation:** Fluctuating smell loss may occur in conductive or obstructive causes due to inflammation. This phenomena is seen in less than 50% of patients.[3]

- **How it was first noticed:** Patient may tell that he misses flavour of food, smell of flowers, sometimes a dangerous gas leak is missed by the patient.

- **Unilateral or bilateral:** Unilateral lesions are more likely to be because of neural origin, although local nasal pathology may produce it. While bilateral lesions are more likely because of conductive lesions like URTIs.

- **Allergies and rhinitis:** Inflammation is the most common cause of anosmia. It may be because of URTIs.

- **Taste:** Often a patient of anosmia will complain of lack of taste or decrease in it. This is due to lack of flavour perception, which is impossible without preserved sense of smell.

- **Trauma:** If present, patient generally reports anosmia since the time of injury, though it is not essential. Cribriform plate fractures, and CSF rhinorrhoea are the associated features.
- **Psychiatric symptoms:** Losing a pleased and known smell may cause depression.[4] A patient of olfactory groove meningioma may have frontal lobe symptoms and memory loss.
- **Symptoms of raised ICP:** One should ask for headache, visual loss, vomiting and look for signs of raised ICP.
- **Associated symptoms:** Visual loss (craniopharyngioma, Foster Kennedy syndrome), epilepsy (temporal lobe lesions).
- **Past history:** Meningitis, renal failure, metabolic problems (thiamine deficiency, B_{12} deficiency, adrenal and thyroid diseases, cirrhosis, etc.) may be the cause. Medical problems like Alzheimer's disease, Parkinson's disease, Sjögren's syndrome needs to be evaluated.
- **Drugs:** Chemotherapy and radiotherapy to the head and neck may affect the sense of smell. Other drugs to affect smell are amphetamines, phenothiazines, prolonged use of nasal decongestants, oestrogen, levodopa, amiodarone. Zinc gluconate spray (used for common cold treatment) can cause permanent anosmia.[5]
- **Personal history:** Chronic smoking, toxins like cadmium may lead to anosmia
- **Surgery:** Cranial and nasal surgery may lead to anosmia.
- **Family history:** Important syndromes that may be associated with anosmia are Turner's syndrome and Kallman's syndrome.

Examination

- **Inspection:** Look for changes of rhinitis: Alae redness and edema; nasal passages for presence of polyps. Ensure they are open before progressing further with the examination. Deviated nasal septum should not be the sole reason for anosmia. Nasal or sinus surgeries may have left the incision scars.

  Anterior rhinoscopy can be used to assess visualisation of the anterior nasal passages to assess for rhinitis, nasal polyps, etc. However, it is unable to visualise olfactory clefts, thus ruling out conductive olfactory loss with it is difficult.

Flexible endoscopy to visualise olfactory cleft, middle meatus, and sphenoethmoidal recess, provides better information. One should look for polyps, fractures, rhinorrhoea, inflammatory edema, extensive crusting with friable and bleeding walls (seen in Wegner's granulomatosis), yellow submucosal nodules (sarcoidosis), and tumours.

- **Tests of olfaction:** Olfaction should be examined in individual nostril. Abnormal expected side should be examined first. Do not use irritants, otherwise Vth cranial nerve gets stimulated. Close the eyes of the patient. Ask him to blow on a metal speculum to see for bilateral frosting.

  With the eyes closed, present a known substance like tea, ginger, soap, etc. to the patient and ask to sniff, and then whether he can smell, and identify them. Females have more ability to identify as compared to males. If the person can smell, his olfactory apparatus is meant to be fine. Repeat this on other side.

- **University of Pennsylvania smell identification test (UPSIT):** It consists of 40 microencapsulated odorants. Patient has to identify them from a choice of four. One with hyposmia can identify 20–34 of them, and one with anosmia can identify 7–19 of them correctly.

- **Sniffin' sticks:** It consists of odour identification (16 common pairs), odour discrimination (16 pairs), and threshold tests (n-butanol).[6] Odours are presented in felt-tip pins.

- **Sniff magnitude test (SMT)** (CompuSniff, LLC, Cincinnati, Ohio): It is based on reduction of sniffing which occurs with the presentation of an odour. Patient wears a cannula connected to a transducer that is connected to a processing board, the result of which is shown on laptop. Patient is presented with either no odour or 5 ml of odour stimulus diluted in mineral oil. Participant's sniff depends on whether he can smell it or not. The transducer senses the sniff's negative pressure. This test has minimal dependence on language and cognitive abilities.[7]

- **MMSE:** To rule out cranial causes.

- **Cranial nerves:** After olfactory nerve, all other cranial nerves, especially the second should be examined to rule out intracranial causes. Second nerve can be compressed by large frontal lobe and olfactory groove masses.

Investigations: Generally, these are not indicated, and should be done in only specific suspected situations.

1. **CT head:** It is indicated when an intracranial pathology is suspected, though a patient of it will present with other neurological symptoms and signs. It will show intracranial SOLs, olfactory cleft abnormalities, and inflammations.

2. **CT PNS:** It is indicated when a paranasal sinus abnormality is suspected, like tumours, inflammations, etc.

3. **MRI brain:** Whenever an intracranial pathology is found on CT head, or neurodegenerative disorder is suspected. Figure 5.1 shows an olfactory groove meningioma with axial and sagittal images.

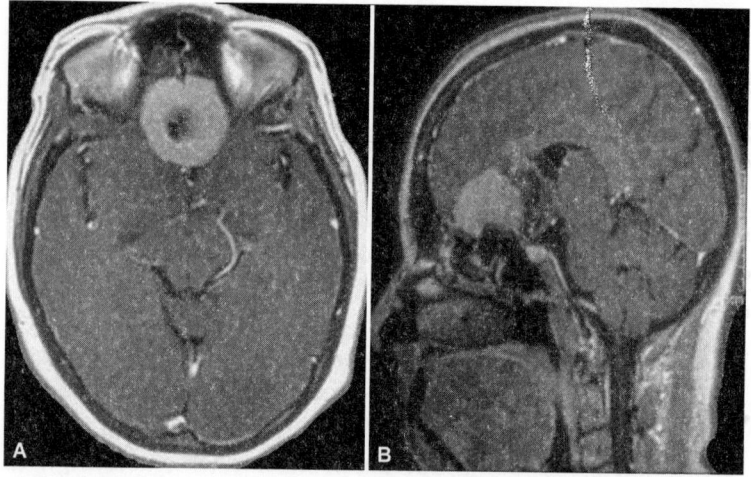

Fig. 5.1A and B: (A) Axial and (B) Sagittal image of an olfactory groove meningioma presenting with headache, memory loss, and anosmia.

4. **Nasal biopsy:** Indicated when sinonasal masses are found, Wegner's granulomatosis is suspected (granulomatous inflammation with vasculitis), or sarcoidosis is suspected (non-caseating granuloma).

5. **Serum vitamin B_{12}, copper, zinc:** These are indicated when these vitamin and mineral deficiencies are suspected.

6. **Serum C-ANCA antibodies:** This is positive in Wegner's granulomatosis

7. **Serum SS-A and SS-B antibodies:** These are positive in Sjögren's syndrome.

8. **Schirmer test:** This is positive in Sjögren's syndrome.

9. **Karyotyping:** When Turner's syndrome is suspected.

Differential diagnosis: Broadly divided into conductive and sensorineural causes.

Conductive[2]

- Common cold
- Allergic and non-allergic rhinitis
- Sinusitis
- Fractured cribriform plate (head injury), ethmoid bone fracture
- Nasal polyps
- Nasal masses

Sensorineural

- Ageing
- Neurodegenerative diseases like Alzheimer's disease, Parkinsonism
- Diabetes
- Toxins
- **Drugs:** Chemotherapeutics, cadmium toxicity, antihistamines, propylthiouracil, antibiotics, levodopa, cocaine, amphetamines, amiodarone
- Hypothyroidism, diabetes, hypoglycaemia, adrenal deficiency, cirrhosis, renal failure
- Neoplasms: Olfactory groove meningioma, frontal lobe masses, temporal lobe masses, sellar/parasellar masses, craniopharyngiomas, neuro-olfactory tumour (esthesio-neuroblastoma)
- Foster-Kennedy syndrome: It is described in frontal lobe masses. Components of it are ipsilateral optic atrophy, contralateral papilledema, ipsilateral central scotoma, and ipsilateral anosmia. Pseudo-Foster-Kennedy syndrome is described when there is absence of mass.
- Vitamin deficiency (B_6, B_{12}, A), zinc or copper deficiency
- Chronic alcoholism
- Radiation therapy
- Stroke, epilepsy

- Ciliopathy due to primary ciliary dyskinesia (Kartagener's syndrome, Afzelius' syndrome or Siewert's syndrome),[7] Parkinson's disease, Alzheimer's disease.

REFERENCES

1. Harris R, Davidson TM, Murphy C, Gilbert PE, Chen M. Clinical evaluation and symptoms of chemosensory impairment: one thousand consecutive cases from the Nasal Dysfunction Clinic in San Diego. *Am J Rhinol*. 2006;20(1):101–108.
2. Seiden AM, Duncan HJ. The diagnosis of a conductive olfactory loss. *Laryngoscope*. 2001;111(1):9–14. doi:10.1097/00005537-200101000-00002.
3. Heald C. Sense and scent ability. BBC News. http://news.bbc.co.uk/2/hi/uk_news/magazine/6199605.stm. Published 2006. Accessed March 19, 2017.
4. Jafek BW, Linschoten MR, Murrow BW. Anosmia after intranasal zinc gluconate use. *Am J Rhinol*. 2004;18(3):137–141.
5. Hummel T, Sekinger B, Wolf SR, Pauli E, Kobal G. "Sniffin" sticks': Olfactory performance assessed by the combined testing of odor identification, odor discrimination and olfactory threshold. *Chem Senses*. 1997;22(1):39–52.
6. Frank RA, Gesteland RC, Bailie J, Rybalsky K, Seiden A, Dulay MF. Characterization of the sniff magnitude test. *Arch Otolaryngol Head Neck Surg*. 2006;132(5):532–536. doi:10.1001/archotol.132.5.532.
7. Ul Hassan A, Hassan G, Khan SH, Rasool Z, Abida A. Ciliopathy with special emphasis on Kartagener's syndrome. *Int J Health Sci (Qassim)*. 2009;3(1):65–69.

Visual Disturbance

```
Visual disturbance
    │
    ├─────────────────────────┐
History                    Examination
  ├ Side                      ├ General
  ├ Onset, duration           ├ Ocular
  │ and progression           └ Neurologic
  ├ How it was first noticed
  ├ Decreased acuity
  ├ Blurring
  ├ Frequent change of glasses
  ├ Diplopia
  ├ Field
  ├ Diurnal variation
  ├ Painful/painless
  ├ Associated symptoms
  ├ Colour perception
  ├ Postural variation
  ├ Trauma
  ├ Drugs
  ├ Malnutrition
  └ Metabolic diseases
```

INTRODUCTION

Vision loss may occur due to either of ocular, neurologic, or systemic cause. Through history and examination, we will try to get to the probable diagnosis.

History

- **Side:** Determine which side is involved. Left/right or both. If unilateral visual disturbance is found search should be made for local cause, while if bilateral disturbance is found both local and systemic causes are possibilities.
- **Onset:** Insidious onset may occur in papilledema, cataract, or raised ICP; while sudden onset may occur in CVA, vitreous haemorrhage, trauma, pituitary apoplexy.
- **Duration:** Short duration is because of raised ICP, intermediate duration may be because of glaucoma, long duration may be because of cataract.
- **Progression:** Enquire which eye is involved first. There may be temporal sequence of involvement of eyes.
- **How it was first noticed:** Enquire how the vision loss was first noticed—while reading, watching television, looking for distant/near objects, blurring of vision.
- **Decreased acuity:** Enquire whether patient is finding difficulty in reading distant, near or both. Distant vision—myopia, near vision—presbyopia, hyperopia; Both—raised ICP, optic nerve pathologies.
- **Blurring:** It may occur in cataract, raised ICP, optic nerve pathologies
- **Frequent change of glasses:** Vision improving with change of glasses is a feature of refractive errors.
- **Diplopia:** It may occur with IIIrd, IVth, VIth cranial nerve palsies, as well as extraocular muscle palsies.
- **Field:** Enquire whether patient is having visual loss in central, peripheral or both. In peripheral field loss, patient may complain of banging with doors/side walls/an object kept sideways while walking; he may not be able to see vehicles coming from sides. Central visual loss occurs in macular abnormalities like diabetic eye, while peripheral field loss occurs due to compression along the optic pathway. Enlarged blind spot is a feature of papilledema.
- **Diurnal variation:** Enquire whether there is diminution in day or night. Diminution in day is known as **Hemeralopia**, also called day blindness. Diminution in night is known as **Nyctalopia** also called night blindness. **Hemeralopia** is different from photophobia in which eye discomfort or pain occurs in light. Causes of hemeralopia are drugs like

trimethadione, Adie'e pupil (fails to constrict in light), aniridia, albinism, central cataracts, cancer associated retinopathy (CAR), Cohen syndrome (obesity, mental retardation, craniofacial dysmorphism), pigmentary chorioretinitis, optic atrophy or retinal/iris coloboma, unilateral or bilateral postchiasmatic brain injury (may also cause night blindness). Causes of night blindness (nyctalopia) are vitamin A deficiency, retinitis pigmentosa (most common cause), pathological myopia.

- **Painful/painless:** Painless loss may occur in ischemia, painful loss may occur in trauma, inflammation, infection, angle closure glaucoma (sudden painful loss of vision).
- **Associated symptoms** should be looked, for example, proptosis, headache, watering, redness, photophobia.
- **Difficulty for colour:** It may be total or partial (red-green, blue-yellow). Colour differentiation is lost in genetic diseases like X-linked disorders, retinitis pigmentosa (in late stages). It may also be acquired due to trauma or pathology of optic nerve. In optic nerve disease red is lost first, and violet is lost last. Pathologies of optic chiasm and posterior to it in the optic pathway does not lead to colour vision loss.
- **Postural variation:** May suggest raised ICP.
- **Trauma history** should always be asked
- **Drugs:** Eye drops containing steroids, ethambutol, trimethadione, amiodarone.[1]
- **Malnutrition:** Vitamin A deficiency
- **Metabolic diseases:** History of hypertension, diabetes, hyperlipidemia should be asked.

Examination

- **General:** Look for hypertension, signs of atherosclerosis.
- **Ocular:** Examine the anterior chamber; see for lens opacities, cataract, iris abnormalities.
- Neurologic
  - Higher mental functions: Examine higher mental functions in suspected cases
  - Examination of IInd cranial nerve:
- **Optotype acuity:** Distant—it is examined with Snellen's chart. Near—with Jaeger's chart. For distant acuity, patient is asked to stand at a distance of 6 metres. With one eye

closed, he reads the bottommost line. If he cannot, then he progresses upwards. Same is repeated with the other eye. The acuity is labelled as 6/5, 6/6, 6/9, 6/12, 6/24, 6/36 and 6/60. Normal is 6/6; 6/5 is considered supernormal.

The test is then repeated with pin hole, and after wearing spectacles (if any). Improvement with these indicates refractive errors.

- **Vernier acuity:** Here, the disalignment among two line segments is checked, monocularly and binocularly. For example, judging the offset between the two line segments shown in Fig. 6.1. This is a test for cortical processing.

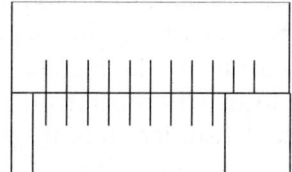

Fig. 6.1: An example of vernier scale

- Grating acuity: It is used in infants and children in which optotype acuity cannot be measured. This can be checked monocularly and binocularly. Child is shown two cards— one plain and one striped. The striped one is initially kept hidden behind the plane one. Then it is moved. The child will follow the striped card. Example of cards is shown in Fig. 6.2.

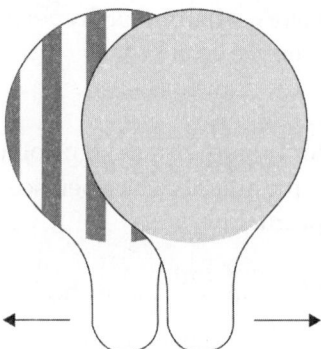

Fig. 6.2: Grating test

- Colour vision: With Ishihara chart.

- Pupils: Examine both pupils. Asymmetry of 1 mm is considered normal. Flash light alternately in both eyes, and look for direct (ipsilateral reaction) and indirect (contralateral) pupillary reaction. Afferent arc is IInd nerve, centre is pretectal nucleus, and efferent arc is formed by IIIrd nerve. Normally, on flashing the light in one eye causes both ipsilateral and contralateral pupil to constrict. Normal and abnormal findings are summarised in Table 6.1. Afferent pupillary defect is seen in retinal or optic nerve pathologies.

Table 6.1: Pupillary reaction			
Pathology	*Flash*	*Direct*	*Indirect*
Rt optic nerve	Right eye	Absent	Absent
	Left eye	Present	Present
Lt optic nerve	Right eye	Present	Present
	Left eye	Absent	Absent
Rt occulomotor	Right eye	Absent	Present
	Left eye	Present	Absent
Lt occulomotor	Right eye	Present	Absent
	Left eye	Absent	Present

- **Accommodation reflex:** It is needed only in cases of abnormal pupillary reflex. Ask the patient to fix his gaze to a distant object. Bring your finger at a distance of 25 cm, and ask him to shift his gaze to it. Notice pupillary constriction and inward movement of both the eyes. The afferent limb is IInd nerve, and efferent is IIIrd nerve. Visual association areas are the control centre.

- **Visual field:** It is tested by confrontation, perimetry (measurement of visual field on a curved surface), campimetry (measurement of visual field on a flat surface).

Confrontation: Sit in front of the patient at a distance of 1 metre at the same eye level. Explain the procedure to the patient. Ask him to close his one eye, and look straight into your eyes. Close your opposite eye. Bring your finger outside inwards in a vibrating manner. Patient has to say 'yes' whenever he sees it. Examine it in all quadrants for each eye.

- Ophthalmoscopy: Direct ophthalmoscopy should be done in every case of visual abnormalities to look for retinal causes, optic atrophy, papilledema, etc. First look red reflex, at a distance of about 50 cm; this rules out any corneal, anterior chamber, or lenticular opacities.[2] Then bring ophthalmoscope closer to the patient, and notice optic disc colour, margins, cup, cup-disc ratio, retinal vessels, lamina cribrosa, and vasculature. Differentiate between primary or secondary optic atrophy based on this. Their findings are summarised in Table 6.2.

Table 6.2: Difference between primary and secondary optic atrophy			
S.No.	Variable	Primary optic atrophy	Secondary optic atrophy
1.	Mechanism	Optic nerve fibers degenerate in an orderly manner with replacement by glial tissue and preservation of optic nerve head	Optic nerve fibers exhibit marked degeneration with excessive proliferation of glial tissue, without preservation of optic nerve head
2.	Disc colour	Chalky white	Grey or dirty grey
3.	Disc margin	Sharply demarcated	Lost and margins are poorly defined
4.	Retinal vessels	Normal	Peripapillary sheathing of arteries as well as tortuous veins
5.	Lamina cribrosa	Well defined	Obscured
6.	Cup	Deep	Elevated
7.	Cup-to-disc ratio (normal 0.1–0.5)	Increased	Decreased
8.	Causes	Optic nerve pathway tumour, pituitary tumour, sphenoid wing meningiomas, tuberculum sellae, meningiomas	Secondary to papilledema or papillitis
9.	Other findings		Hyaline bodies (corpora amylacea), and drusen may be seen

Table 6.3: Clinical features of lesions at different parts of visual pathway[3]

Entity	Visual acuity	Colour vision	Visual field	Pupillary reaction	Optic disc
Macula	Affected	Affected	Central scotoma	Possible mild APD	Normal
Optic nerve	Affected	Affected	Ipsilateral central, paracentral or cecocentral scotoma	APD	Atrophy/ edema*
Distal optic nerve near chiasm	Affected	Affected	Junctional scotoma	APD	Normal/ atrophy
Chiasm	Normal	Normal	Bitemporal hemianopia (congruous)	Normal	Normal
Optic tract	Normal	Normal	Contralateral Incongruous homonymous hemianopia	Mild APD in contra-lateral eye	Normal
LGB	Normal	Normal	Contralateral Incongruous homonymous hemianopia	Normal	Normal
Optic radiations-Temporal lobe	Normal	Normal	Contralateral superior quadrantopia	Normal	Normal
Optic radiations	Normal	Normal inferior	Contralateral quadrantopia	Normal	Normal
Calcarine cortex	Normal	Normal	Contralateral congruous homonymous hemianopia	Normal	Normal

* Edema in papillopathy

Abbreviations—APD: Afferent papillary defect, LGB: Lateral geniculate body

Investigations

- **CT head and orbit:** It is advised in cases of trauma. One can find ocular hematoma, orbital hematoma, orbital fracture, foreign body, bone chip compressing the optic nerve, intracranial space occupying lesions.
- **MRI brain:** It can show intracranial space occupying lesions, multiple sclerosis.
- **CT angiogram brain:** It can show cavernous sinus aneurysm.
- **Lumbar puncture:** It can be done to see for raised opening pressure in suspected of benign intracranial hypertension.
- **Serum investigations:** In appropriate situations, one may order WBC count, syphilis, lyme, tubercular, and ACE serology (sarcoidosis). Low cortisol may be found in pituitary apoplexy.

Differential Diagnosis

1. **Ocular causes:** Vast number of ocular pathologies may affect vision. These include abnormalities of cornea, anterior chamber, lens, posterior chamber, and retina.
2. **Cortical visual impairment:** It is also called cerebral visual impairment, or retrogeniculate visual impairment. Vision loss may occur due to cortical or subcortical injury. Both grating acuity and vernier acuity are reduced; vernier acuity being specific.
3. **Traumatic vision loss:** There will be history of recent trauma. Examination may show decreased visual acuity, and signs of trauma. CT head and orbit is the investigation of choice.
4. **Optic neuritis:** It is predominantly seen in middle aged females. Eye movements may become painful. Prior neurological symptoms like paraesthesias, weakness may be present. Examination will reveal decreased visual acuity, papillitis. MRI brain is the investigation of choice.
5. **Raised ICP:** It can be because of hydrocephalous, venous sinus thrombosis, or space occupying lesion in the brain. Patient may complain of gradually progressive vision loss, diplopia (due to 6th nerve palsy), headache, vomiting, etc. Ophthalmoscopy will reveal features of papilledema. MRI brain with venogram may be ordered to look for above mentioned pathologies.

6. **Space occupying lesion:** Tumours may compress or involve the optic pathway directly. Examples include cranio-pharyngioma, optic pathway gliomas, pituitary lesions, etc. Patient may complain gradual progressive vision loss. MRI brain will be diagnostic.

7. **Pituitary apoplexy:** Patient may complain sudden vision loss, headache, altered consciousness. Examination may reveal decreased visual acuity, ptosis, absent pupillary reflex. Serum cortisol may be low. MRI brain is diagnostic of pituitary hemorrhage.

8. **Transient ischemic attack:** There will be sudden painless mono-ocular vision loss. It may lasts for a few minutes. Hypertension, diabetes, hypertriglyceridemia are precipitating factors. Examination finding between the episodes may be normal. Serology will reveal elevated lipid profile. MR angiography may reveal carotid atherosclerotic disease.

REFERENCES

1. Purvin V, Kawasaki A, Borruat FX. Optic neuropathy in patients using amiodarone. *Archives of ophthalmology (Chicago, Ill. : 1960)*. May 2006;124(5):696–701.

2. Donahue SP, Baker CN. Procedures for the Evaluation of the Visual System by Pediatricians. *Pediatrics*. Jan 2016;137(1).

3. Campbell WW. De Jong's The Neurologic Examination. Seventh ed. Philadelphia: Wolters Kluwer/Lippincott Williams & Wilkins; 2013.

Diplopia

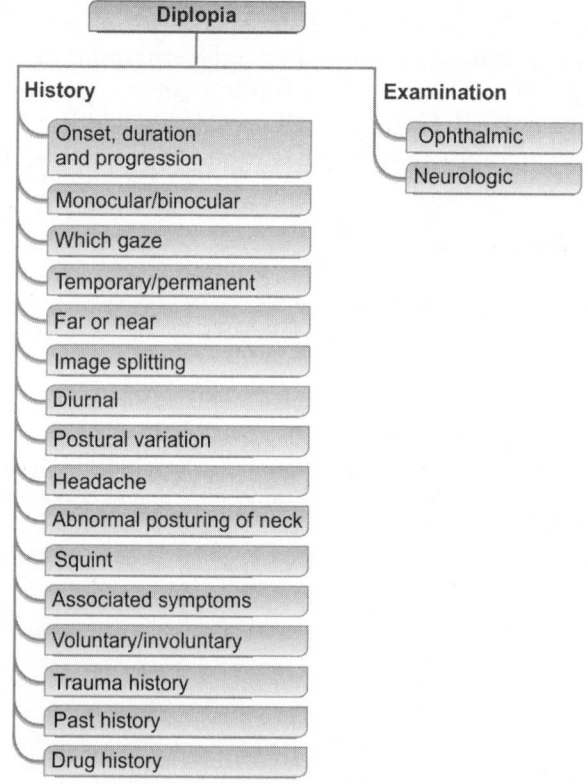

Diplopia

History

- Onset, duration and progression
- Monocular/binocular
- Which gaze
- Temporary/permanent
- Far or near
- Image splitting
- Diurnal
- Postural variation
- Headache
- Abnormal posturing of neck
- Squint
- Associated symptoms
- Voluntary/involuntary
- Trauma history
- Past history
- Drug history

Examination

- Ophthalmic
- Neurologic

Definition

Disorder of vision in which two images of a single object are seen because of unequal action of the eye muscles—also called *double vision*.

History: A comprehensive history is the most useful part of approach to a patient with diplopia.

Onset, duration and progression: Sudden onset may occur in trauma, infection or inflammation, CVA, diabetes (due to vasculitis), and PComm aneurysms. Chronic progressive lesion occurs in tumours, abscesses, multiple sclerosis (MS). Transient recurrent ones occur in epilepsy, ophthalmoplegic migraine.

Monocular/binocular: There is generally binocular diplopia, which is present only when person sees from both eyes, and is disappeared when he closes one eye. Monocular diplopia is very rare and is because of ophthalmic causes like corneal opacities, keratoconus, multiple openings in pupils or iris, cataracts, subluxation of lens, or non-organic lesion.

Which gaze: Asking patient to which gaze he is having diplopia can suggest the cause of diplopia. Diplopia occurs towards the side of action of paretic muscle. For example, IIIrd nerve lesion will have diplopia on looking upwards, downwards or medially. A left VIth nerve lesion will have diplopia on looking towards left side. A IVth nerve lesion will have diplopia on walking downstairs. Questions should be asked for following gazes: Primary, left, right, up, down, up and laterally, down and laterally for both eyes.

Temporary/Permanent: Temporary diplopia can be caused by Alcohol intoxication, concussion injuries, drugs like phenytoin, zonisamide, lamotrigine, zolpidem, ketamine, dextromethorphan, and tired or strained eye muscles.

Far or near vision: Ask the patient whether the diplopia occurs for far or near vision. If it occurs for distant vision, the cause is lateral rectus paresis, while if for near vision, the cause is medial rectus paresis.

Image splitting: Cause of image splitting vertically may be either superior or inferior rectus paresis. Cause of image splitting horizontally may be either lateral or medial rectus paresis. While cause of diagonal splitting of image may be obliqui weakness.

Diurnal: Diplopia occurring at the end of day is generally due to myasthenia gravis.

Postural variation: Diplopia increasing on bending forward may be because of increased ICP.

Pain on eye movements: It is a sign of inflammation and may be because of abscess, sinusitis.

Headache: It may occur because of raised ICP, meningitis.

Abnormal posturing of neck: Patient tilts his neck in the direction of the paretic muscle. For example, in right IIIrd nerve lesion head tilts leftwards, in IVth nerve lesion head tilts downwards and forwards when walking, in right VIth nerve lesions head tilts rightwards.

Squint: Divergent squint occurs in third cranial nerve palsy. Convergent squint occurs in sixth cranial nerve palsy. Fourth cranial palsy rarely has a squint.

Associated symptoms: Ptosis may be because of IIIrd nerve palsy; proptosis may be seen in cavernous sinus lesion, orbital mass.

Voluntary/involuntary: Some patients can do voluntary excess convergence causing diplopia. It is usually a cause of fun and is non-organic.

Trauma history: Recent trauma to face and head may lead to diplopia by causing orbital fractures, blow out fractures with hematoma inside the orbit, entrapment of soft tissue and muscles.

Past history: Enquiry should be made for past history of diabetes, vascular disease, hypertension, headache and other neurologic complaints, muscle fatigue or weakness.

Drug history: Drugs like phenytoin, zonisamide, lamotrigine, zolpidem, ketamine, dextromethorphan, and flouroquinolone antibiotics can cause diplopia.

Examination[1]

We have divided this in ophthalmic (with anatomical and physiological examination) and neurologic examination and presented it in a format which will be easy to remember.

Ophthalmic

Inspection

Head position: Look the neutral position of the head; there may be subtle tilt of head to avoid diplopia. Evaluate the diplopia by asking the patient to tilt his head. A head tilt to the direction of paretic muscle will decrease the diplopia, and vice versa. For example, a right superior oblique muscle weakness will cause head deviation towards left and downwards.

Lid position: Look for lid position of both eyes. Asymmetry of lids may accompany vertical strabismus. Ptosis, if present, may indicate IIIrd nerve palsy. Eyelid retraction is a feature of thyroid ophthalmopathy.

Ocular: Examine to look for signs of inflammation, corneal opacities, lenticular opacities, obvious deviation of eyes, orbital masses, and vascular engorgement (carotico-cavernous fistula). Note for skew deviation of eyes (whether one eye is directed upwards and the other downwards) which can occur in eighth cranial nerve, labyrinth and cerebellum lesions.

Proptosis: Examine to look for proptosis by standing behind the patient (*see* chapter on proptosis).

Red reflex: Examine red reflex through an ophthalmoscope and see for any obstruction like lenticular or corneal opacities.

Light reflex: Both direct and indirect light reflex should be examined for delineating IIIrd nerve palsy (*see* chapter on visual loss). Pupil sparing third nerve palsy (normal light reflex in accompaniment of ptosis, and loss of IIIrd nerve functions) occurs in diabetes. This generally recovers in 6 weeks.

Accommodation reflex: Should be seen, but is of less value if light reflex is normal (*see* chapter on visual loss).

Visual acuity: Far, near and colour vision should be examined to look for defects in visual pathway. Pinhole should also be used for examination. Improvement in diplopia with pinhole suggests intraocular or refractive problems.

Extra-ocular movements: Evaluate for diplopia in primary gaze as well as in nine cardinal gazes of vision. This will point out which muscle or nerve is involved. The nine cardinal positions of gaze are straight, nasal, up and nasal, down and nasal, up, up and temporal, temporal, down and temporal, and down. Evaluate each eye separately, known as ocular ductions, and then combined, known as ocular versions.

Ask the patient to fix his gaze on your finger. Take your finger in a 'H'-shaped manner, i.e. both lateral, up and down in lateral most position bilaterally, primary, up, and down gazes. Figure 7.1 shows the muscles acting for each movement.

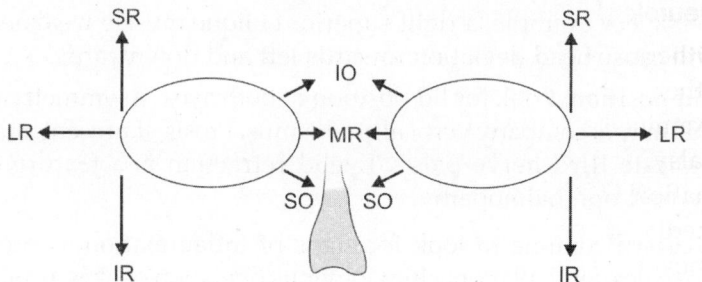

Fig. 7.1: Actions of different extra-ocular muscles have been shown. IO: Inferior oblique; IR: Inferior rectus; LR: Lateral rectus; MR: Medial rectus; SO: Superior oblique: SR: Superior rectus

Cover/uncover tests: This test can be used to find out whether the diplopia is monocular or binocular. Should be done to look for latent squint.

Ask the patient to fix on a near object (33 cm), cover his fixating eye with a black card. The mal-aligned eye may move inward or outward (inward in exotropia and outwards in esotropia). Repeat this test for distance (3 m) and far distance (>3 m).

Peripheral visual fields: Evaluate this by confrontation testing. There may be abnormalities in peripheral visual fields due to intracranial masses compressing visual tracts or cranial nerves (*see* in visual loss chapter to see how to examine).

Palpation

Orbital rim: Palpate it to look for orbital fractures, and any defect. Do this bilaterally.

Between orbital rim and globe: With the eyes closed, palpate between orbital rim and globe of eyes to look for tumours. Do this bilaterally.

Over the eye globe: With the eyes closed, palpate gently over the globe to look for fullness that may be seen in raised ICP. Compare this in both eyes.

Percussion

Percuss over orbital rims, bilaterally, to see for any tenderness.

Auscultation

Auscultate over closed eyes and over thin temporal bone at the outer margin of orbit to look for bruit, which is present in carotico-cavernous fistula.

Neurologic Examination

Other cranial nerves: After second, third, fourth, and sixth, all other cranial nerves especially fifth and seventh should be examined. Osteitis of the petrous apex may lead to abducens palsy, facial pain secondary to gasserion ganglion inflammation, and seventh nerve palsy due to suppurative otitis media—syndrome described by Gradinego in 1907,[2] and is known after him.

Sensorimotor examination: Do a complete sensorimotor examination. In third nerve palsy with hemiplegia and ataxia—suspect midbrain lesions; in sixth and seventh nerve lesions along with motor paralysis—suspect for pontine lesions.

Tensilon test: It is done when diplopia is fluctuating with the fatigue and is more at the end of the day. It is specific for myasthenia gravis.

Patient is given 10 mg of edrophonium chloride (first give test dose of 1 mg to see for any hypersensitivity). A positive response will be immediate resolution of diplopia along with increased salivation, lacrimation, and flushing. Atropine should be ready beforehand, in case of any untoward incidence of bradycardia.

Investigations

Imaging studies: These are indicated only when there is diplopia in a patient of <50 years of age with neurological findings, progressive diplopia, or with a history of cancer.[3]

See old photographs: Compare old photographs with present condition to see for head position, eyelid position, eye deviations, etc. This will help to see for progression of the disease.

CT skull and orbit: This is indicated when we suspect tumours of orbit or intracranial region, orbital wall fractures, raised ICP, aneurysms (with angiogram).

MRI head and orbit: When tumours are diagnosed and operative plan has to be decided. Sinusitis is also better seen on MRI.

Angiography: This is indicated when carotico-cavernous fistula is suspected.

Differential Diagnosis[4]

Third nerve palsy

Nuclear lesions (usually bilateral) and midbrain (associated with hemiplegia or ataxia or tremors): Tumours, haemorrhage, MS, CVA.

Interpeduncular space (isolated IIIrd nerve palsy): Aneurysm, neurosyphilis, TBM, traumatic fracture base of skull.

Cavernous sinus (associated with IVth, ophthalmic, maxillary and VIth palsy): Aneurysm, thrombosis, fistula.

Orbit (associated with IInd, IVth, ophthalmic, and VIth palsy): Tumour, fracture, aneurysm.

Fourth nerve palsy

Nuclear lesions and midbrain (associated with IIIrd nerve palsy): Tumours, MS, CVA

Interpeduncular space (isolated IVth palsy): Aneurysms, meningitis, fracture

Cavernous sinus (associated with IIIrd, ophthalmic, maxillary and VIth palsy): Aneurysm, thrombosis, fistula

Orbital (associated with IInd, IIIrd, ophthalmic, and VIth palsy): Mass, fracture, aneurysm.

Sixth nerve palsy

Nuclear lesions and pons (associated with VIIth nerve palsy): Tumours, CVA, MS, raised ICP (false localizing sign)

Cavernous sinus lesion (along with IIIrd, IVth nerve, ophthalmic, maxillary): Tumours, aneurysm, thrombosis, fistula.

Orbital (along with IInd, IIIrd, IVth): Fracture, neoplasm, aneurysm.

REFERENCES

1. Campbell WW. *DeJong's THE Neurologic Examination*. Seventh. Philadelphia: Lippincott Williams & Wilkins; 2013.
2. Gradinego G. Uber die paralyse des n. Abduzens bei otitis. *Arch F Ohrenheilk*. 1907;74:149–158.
3. Murchison AP, Gilbert ME, Savino PJ. Neuroimaging and acute ocular motor mononeuropathies: a prospective study. *Arch Ophthalmol (Chicago, Ill 1960)*. 2011;129(3):301–305. doi:10.1001/archophthalmol.2011.25.
4. Gami N. *Bedside Approach to Clinical Neurology*. 2nd ed. Jaypee; 1998.

Ptosis

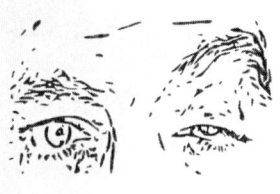

Ptosis

History

Age
- Congenital
- Acquired

Onset, duration, progression
- Sudden
- Insidious
- Chronic

Side

Partial complete

Waxing or waning

Diurnal variation

Diplopia

Redness, excessive lacrimation, discharge

Headache

Association
- Anhidrosis
- Enophthalmos

Relation with mastication

Relation with eyeball movements

Drugs, toxins

Trauma

Diabetes

Wearing contact lens for a long time

Family history

Examination

Eyelids

Pupils

Palpebral fissure height

Marginal reflex distance

Upper eyelid skin crease

Levator function

Higher mental functions

Cranial nerves

Sensorimotor examination

Sympathetic examination

Definition: Drooping or inferior displacement of the upper eyelid.

Anatomy: Opening of the upper eyelid is done by levator palpabrae superioris (LPS) muscle (supplied by IIIrd CN) and superior tarsal muscle (also known as Müller's muscle—supplied by sympathetics). Eyelids are kept open by tonic contraction of LPS. Here, Müller's muscle contributes only 1–2 mm of the opening.[1] Retraction of lower eyelid is done by capsulopalpebral fascia and inferior tarsal muscle (supplied by sympathetic). Closure of the eyelids is done by orbicularis oculi (supplied by VIIth CN). The reflexive blinking of the eyelids is done by coordination of nuclei of IIIrd, Vth, VIIth nerve, and circuits of cerebrum, basal ganglia, and hypothalamus.[2]

History

- **Age of the patient:** Congenital—it may be congenital and can be inherited, e.g. Tay-Sachs disease, amaurotic idiocy. It is generally bilateral, and upper lids are smooth.

  Acquired—most of the cases are acquired. Most common cause is aponeurotic damage of levator palpabrae superioris (because of senescence, infection, inflammation, trauma, chronic contact lens use), or its denervation because of IIIrd nerve palsy.

- **Onset, duration, progression:** Sudden onset may be because of trauma, CVA or aneurysms. Insidious onset may be because of infections (encephalitis, diphtheria, myasthenia gravis, etc.), and chronic ones may be because of congenital, myopathies, syphilis, midbrain tumours, or unruptured aneurysms.

- **Unilateral or bilateral:** Bilateral ones are because of congenital causes (Tay-Sachs, amaurotic idiocy, congenital diplegia), syphilis, encephalitis, myasthenia gravis, or ocular myopathies. Unilateral ones are due to causes like CVA, inflammatory, infective, tumours, aneurysms.

- **Partial or complete:** Complete ptosis is because of IIIrd nerve palsy, while partial ptosis may be because of IIIrd nerve paresis or sympathetic paresis.

- **Waxing and waning:** Fluctuating ptosis is suggestive of myasthenia gravis (MG) in which the ptosis gets worsened as the day progresses and as the patient fatigues. It gets improved with rest.

- **Diurnal variation:** Ptosis increasing as the day progresses is because of MG.

- **Diplopia:** Diplopia during medial gaze suggests IIIrd nerve palsy. In complete ptosis diplopia is absent.

- **Redness, excessive lacrimation, discharge:** These are seen in orbital inflammatory causes.

- **Vision abnormalities:** Vision disturbances should be asked. Orbital pathologies, metabolic disturbances like diabetes, and intracranial pathologies may affect the vision.

- **Headache:** Space occupying lesion may cause headache. Similarly, aneurysmal rupture, cavernous sinus lesion may lead to headache.

- **Associations:** Anhidrosis, enophthalmos may point towards Horner's syndrome due to sympathetic paralysis.

- **Relation with mastication:** This should be asked in cases of congenital ptosis where there may be synkinesia of LPS and mandibular nerve. The eyelid retracts during mastication, sucking, or eating, known as Marcus Gunn jaw-winking syndrome.

- **Relation with eyeball movements:** History should be asked whether there is any association with eyeball movements. In congenital ptosis, due to Duane's syndrome, there may be paresis of adduction, abduction, or both, which may be associated with ptosis and globe retraction.

- **Drugs, toxins:** Drugs like high dose of opioids (morphine, oxycodone, heroin, hydrocodone), pregabalin; toxins like snake venom (black mamba) may cause ptosis.

- **Trauma:** Trauma to the LPS, IIIrd nerve, or sympathetic chain may cause ptosis.

- **Diabetes:** Diabetic neuropathy may be a cause of ptosis.

- **Wearing contact lens for a long time:** It is said to be a possible cause.

- **Family history:** This should be enquired to rule out familial causes.

Examination

- **Eyelids covering cornea/pupils**: Normally upper eyelid crosses the limbus by 1–2 mm, and lower eyelid just touches the limbus.
- **Pupils and their reaction**—Pupils should be examined to find out the IIIrd nerve palsy.
- Patient should be asked to lift the eyelids, after fixing the frontalis muscle with a finger to avoid taking help of it.
- **Palpebral fissure height**: Bilaterally, measure the widest distance between the upper and lower eyelid margins in primary gaze. It is normally 8–11 mm.
- **Marginal reflex distance:** It is the most effective measurement of ptosis, and is independent of lower eyelid position. Patient should fixate in a primary gaze at a distant target. Measure the distance between papillary reflex and upper eyelid, that is normally 4–5 mm.
- **Upper eyelid skin crease:** Distance of the upper eyelid skin crease from the lid margin should be measured, which is normally 8 mm in men and 9–10 mm in women. This is absent in congenital ptosis, and increased in aponeurotic ones.
- **Levator function:** It can be checked by measuring upper lid movement from full closure to its opening just before the frontalis contraction. Examiner can manually stabilise the frontalis muscle to prevent its action. Normally it is 10–12 mm. Above 11 mm is considered very good, 8–10 mm is considered good, 5–7 mm as fair, and less than 4 mm indicates poor function.
- **Higher mental functions:** If intracranial lesion is suspected.
- **Cranial nerves:** All cranial nerves should be examined. IIIrd cranial palsy is a common cause of ptosis. Lesions at cavernous sinuses may lead to ptosis along with Vth and VIth nerve involvements. Brainstem lesions may lead to ptosis, along with multiple cranial nerve palsies, and long tract signs. Nuclear lesions can even cause bilateral ptosis.
- **Sensorimotor examination:** To rule out congenital ptosis in which paralysis of limbs occur. Brainstem lesions can present with ptosis along with long tract signs.
- **Sympathetic examination:** It should be done to rule out Horner's syndrome. See for size of pupils, perspiration over both sides of face, and ciliospinal reflex.

Investigations

1. **MRI brain and orbit:** These are ordered when an intracranial or orbital pathology is suspected. Even if an orbital tumour seems to be confined in orbit, screening of brain should be done to see the extension.

2. **MRI thorax:** If central Horner's syndrome (first order neuron) is suspected and MRI brain is normal, then MRI thorax is advised to rule out tumours at thoracic inlet.

3. **Serum tests:** Levels of glucose for diabetes, lactate and pyruvate for mitochondrial disorders like chronic progressive external ophthalmoplegia (CPEO), acetylcholine receptor antibodies for myasthenia gravis, thyroid function tests can be ordered when in suspicion.

4. **Muscle biopsy:** It is done in cases where myogenic causes of ptosis like CPEO is suspected. Biopsy shows ragged red fibres in gomori-trichome staining in 50% of cases.

5. **Tensilon test: Edrophonium** test can be used in cases of myasthenia gravis (explained in detail in diplopia chapter).

6. **Cocaine (4/10%) or apraclonidine (0.5/1%) test:** These are done when Horner's syndrome is suspected. Cocaine causes mydriasis in the normal eye, but has no effect in the miotic pupil of Horner's syndrome. Apraclonidine has dilatory effect over the miotic pupil due to denervation sensitivity of the iris dilator, and has no effect over the normal pupil.

7. **Hydroxyamphetamine test:** This test is used to differentiate between preganglionic and postganglionic Horner's syndrome. 1% topical hydroxyamphetamine is instilled in the miotic eye. If the pathology is preganglionic, only then the pupil dilates.

8. **Anti-GQ1B antibodies:** These are ordered when there is suspicion of Miller Fisher syndrome.

Differential Diagnosis

- **True ptosis:** On severity basis, ptosis can be classified as minimal (1–2 mm), moderate (2–4 mm), or severe (>4 mm). **On etiology basis, true ptosis can be classified into congenital and acquired ones.**

- **Congenital:** Causes include developmental anomaly of the LPS. Upper eyelid crease is usually absent. The ptosis is less in down-gaze due to the fibrotic levator muscle limits inferior mobility of the eyelid. It is unilateral in 69% of cases.[3] Other causes include Marcus Gunn jaw-winking syndrome in which there is aberrant synkinesis of LPS and mandibular nerve fibres leading to ptosis at rest, and retraction of eyelids during times of chewing and eating. Duane's syndrome is another cause in which there is ptosis along with paresis of adduction or abduction of the eyeball.

- **Acquired:** Causes include mechanical, myogenic, neurogenic, neuromuscular, and cerebral. **Mechanical ones** are the one in which LPS is normal, e.g. involutional/senile ones, in which there is gradual stretching or dehiscence of levator aponeurosis.[4] Other mechanical causes include tumours of orbit, lid edema, trauma, and infiltration.

  Myogenic are the ones in which orbicularis oculi and LPS are involved. Examples include **chronic progressives external ophthalmoplegia**, in which, there is both ptosis and ocular immobility, though ptosis precedes the ocular immobility.[5] Pupils and corneal sensations are spared. Another cause of myogenic one includes autosomal dominant **oculopharyngeal muscular dystrophy**, in which there is ptosis along with extraocular ophthalmoparesis, progressive dysphagia, and proximal limb weakness. In **myotonic dystrophy** there is ptosis along with cataract and pigmentary findings.

  Neurogenic ones include Horner's syndrome, IIIrd nerve palsy, Miller Fisher syndrome, and ophthalmoplegic migraine. In **Horner's syndrome** there is sympathetic paralysis which causes denervation of Müller's muscle. There is lesser degree of ptosis. There is meiosis, anhydrosis, enophthalmos, and is always unilateral. It may be because of lesions at brainstem (thrombosis of PICA, pontine tumours, encephalitis), cervical and thoracic cord (C8-T2 segments), carotid sympathetic plexus (ICA aneurysms, tumour of Gasserian ganglion), thoracic inlet (Pancoast tumour), neck (trauma, surgery, thyroid tumours), or mediastinum (aortic aneurysm, cardiac aneurysm, mediastinal tumours). It is tested with cocaine, apraclonidine, and hydroxyamphetamine tests, as explained above. **IIIrd cranial nerve paralysis** may be because of lesion anywhere from its nucleus in midbrain to orbit. Causes may include tumours, trauma, infections, inflammations, demyelination, and vascular lesions. It may be complete or incomplete IIIrd nerve palsy. **Miller Fisher syndrome (MFS)** is a condition having triad of ophthalmoplegia, ataxia, and areflexia. Its a variant of Guillain-Barré syndrome. Pupil sparing third nerve palsy may occur.[6] **Ophthalmoplegic migraine** is a rare condition with co-occurrence of headache with IIIrd cranial nerve palsy. Recurrent isolated ptosis may also occur, and recurrent demyelination is considered as the presumed etiology.[7] Pupil sparing third nerve palsy may also occur in **diabetes mellitus.**

Neuromuscular ones include **myasthenia gravis.** There is diurnal variability of the symptoms. Levator function is often abnormal and there is unilateral or bilateral ptosis. **Botulism** either by toxin or iatrogenic injection may cause ptosis accompanied with dysarthria, dysphagia, blurred vision or muscle weakness.

Cerebral ptosis may be due to involvement of supraseg-mental upper motor neuron pathways for eyelids. There may be unilateral or bilateral ptosis, with predominance ipsilateral to the hemiparetic side. Causes may include stroke, demyelination, neoplasia, and infections.

- **Pseudoptosis:** In this, the eyelid looks ptotic but is actually not. Conditions causing it include hemifacial spasm, narrow palpebral fissure, contralateral hypertropia, ipsilateral hypotropia, enophthalmos, micro-ophthalmos, anophthalmos, dermatochalasis, contralateral eyelid retraction (as in Graves' disease), or high myopia, and blepharospasm.

REFERENCES

1. Beard C. Muller's Superior Tarsal Muscle: Anatomy, Physiology, and Clinical Significance. *Ann Plast Surg.* 1985;14(4):324–333.

2. Yadegari S. Approach to a patient with blepharoptosis. *Neurol Sci.* 2016;37(10):1589–1596.

3. SooHoo JR, Davies BW, Allard FD, Durairaj VD. Congenital ptosis. *Surv Ophthalmol.* 2014;59(5):483–492. doi:10.1016/j.survophthal.2014.01.005.

4. Dortzbach RK, Sutula FC. Involutional blepharoptosis: a histopathological study. *Arch Ophthalmol.* 1980;98(11):2045–2049.

5. Lee AG, Brazis PW. Chronic progressive external ophthalmoplegia. *Curr Neurol Neurosci Rep.* 2002;2(5):413–417. doi:10.1007/s11910-002-0067-5.

6. Rittenbach TL. Article 4 A Case Presentation of a Third-Nerve Palsy as a Characteristic of Miller Fisher Syndrome.

7. Gelfand AA, Gelfand JM, Prabakhar P, Goadsby PJ. Ophthalmoplegic "migraine" or recurrent ophthalmoplegic cranial neuropathy: new cases and a systematic review. *J Child Neurol.* 2012;27(6):759–766.

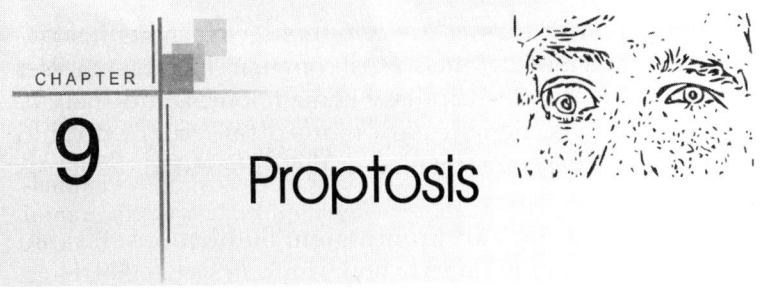

Proptosis

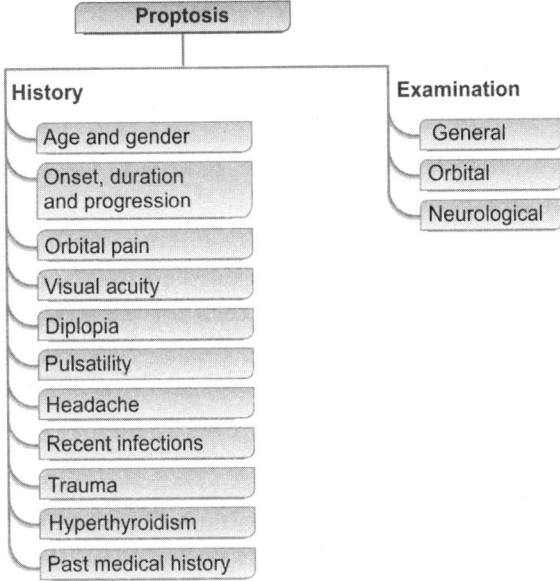

Proptosis can be a result of generalised, orbital and neurological causes. We will limit our discussion to neurological and orbital causes only. One should also keep thinking in the mind the various spaces which can be involved by a mass. They are: The intraconal space (inside the muscles), the extraocular muscles, the extraconal space (outside the extraocular muscles, the subperiosteal space, Tenon's space, and the extraorbital space.

History

- **Age and sex:** In adults, most common cause of unilateral or bilateral proptosis is thyroid ophthalmopathy. Most common adult masses are lymphoid, cavernous heman-giomas, and meningiomas. Most common childhood masses are: Dermoid cysts, capillary hemangiomas, and rhabdo-myosarcomas. Females are more prone to get thyroid diseases.[1] These are detailed in the differential diagnosis section.

- **Onset, duration, and progression:** Sudden onset, rapid progression over minutes to hours indicates a haemorrhage. Processes occurring over hours to days indicate inflam-mation or infection. Gradual progression over weeks to months suggests chronic inflammation, or a malignant neoplasm. While chronic and slow progression over months or years suggest a benign neoplasm.

- **Orbital pain:** Pain typically occurs either in rapid pro-gression of the proptosis, or an inflammatory cause. Severe proptosis may also lead to pain.

- **Visual acuity:** Visual acuity remains unchanged generally till late in the course. If it decreases early, then it means a mass near the optic nerve.

- **Diplopia:** It indicates an extraconal mass (outside the muscles) pressing the eyeball in non-axial direction.

- **Pulsatility:** It indicates a vascular pathology like carotico-cavernous fistula.

- **Headache:** Headache, and other symptoms of raised ICP should be enquired. Sphenoid wing meningiomas, etc. may lead to proptosis. Similarly, **change in coughing/sneezing** suggests a direct communication with intracranial cavity.

- **Recent infections:** Respiratory tract infection etc. in the recent past should be enquired. Sometimes, a temporal relation of proptosis with infection at other sites can be established, which points towards orbital cellulitis as a cause of proptosis.

- **Trauma:** It should be enquired as fractures of orbit, maxilla, frontal, lacrimal bones etc. can lead to proptosis. Post-traumatic vascular anomalies like carotico-cavernous fistula can also be a differential.

- **Symptoms of hypo and hyperthyroidism:** These should be asked for to rule out thyroid pathologies.

- **Past medical history:** As obvious, history of thyroid disease has to be asked as it is the most common cause of proptosis in adults. History of lymphoproliferative diseases, neo-plasm, etc. in other body parts should be enquired.

Examination: It should be divided into general examination, orbital examination and neurological examination.

General examination: Vitals and other signs should be seen for hyper and hypothyroidism.

Orbital examination: Inspection, palpation, auscultation

- **Acuity:** It should be examined as early loss of it suggests pathology near the optic nerve.
- **Fundus:** Optic atrophy if present suggests pathology of or near the optic nerve.
- **Proptosis:** (from lateral canthus to corneal apex): If more than 20 mm, then it is called proptosis. If the difference between two eyes is more than 2 mm, then it is significant. It is measured by Hertel's ophthal-mometer as shown in Fig. 9.1. Measurement

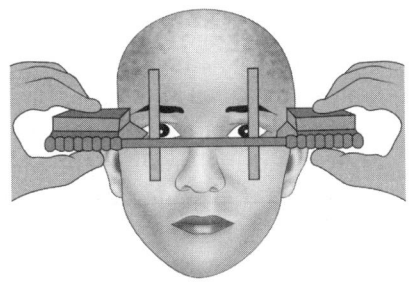

Fig. 9.1: Hertel's ophthalmometer

is taken from the lateral orbital rim to the corneal apex. Other methods are Naugle (from superior orbital and inferior orbital margin, thus can be used in lateral orbital rim fractures), and Luedde method (fixes the ruler on the lateral orbital wall).
- **Extraocular movements:** Check extraocular movements. These may be hindered due to mechanical causes.
- **Periocular changes:** Look for changes in skin, conjunctiva, and periocular tissues. Lid lag is seen in thyroid ophthal-mopathy; temporal flare is also seen in it. Conjunctival salmon patch may be seen in orbital lymphoma. Congestion suggests cellulitis, raised ICP, or tumour impairing the venous return.
- **Palpate between orbit and globe:** Palpate here to see for any mass lesion.
- **Orbital bruit:** Auscultate over closed eye and thin temporal margin of orbital bone to look for it. If present, suggests carotico-cavernous fistula.

- **Mastoid bruit:** Auscultate over mastoid. If bruit is found, it suggests communication between cavernous sinus and inferior petrosal vein.
- **Valsalva manoeuvre:** Enlargement of the lesion suggest venous lesion.

Neurological Examination

- Higher mental functions: This should be examined if any intracranial mass is suspected along with proptosis.
- Cranial nerves: IInd, IIIrd, IVth, VIth should be examined in detail.

Investigations

- **CT orbit:** Apart from clear cases of Graves' disease, orbital CT should be advised. It will reveal fractures, mass lesions, compartment of the mass, intracranial extension, etc. Contrast enhancement will be seen in inflammatory, vascular, meningiomas, malignant masses. Angiography can be combined if vascular lesion is suspected. One should watch whether the mass is a bone eater (malignant), or pusher (benign).
- **MRI:** Its main indication is when mass is involving orbitocranial junction, or optic chiasm. Vascular lesions, cystic lesions are also better seen than CT. Location of optic nerve is also seen better in it, which will guide us for planning approach for surgery.
- **DSA:** It is required when vascular lesions like caroticocavernous fistula is suspected.
- **Thyroid profile:** It should be done to rule out Graves' disease.

Characteristics

- **Intraconal mass:** Early visual loss, impairment of ocular motility, axial proptosis.
- **Extraconal:** Proptosis early, nonaxial
- **Intracanalicular:** Early visual loss, papilledema, appearance of optociliary shunt vessels on surface of optic discs, minimal or no proptosis.

Differential Diagnosis

- **Inflammatory:** Graves' disease, tuberculoma
- **Infectious:** Orbital cellulitis, cysticercosis

- **Haemorrhagic**
- **Neoplastic:** Optic nerve glioma, meningioma, dermoid cyst, rhabdomyosarcoma, lymphoma, hemangioma, neurofibromas (most commonly from ciliary nerves), lipomas, bony tumours of the orbit, lacrimal gland masses—adenoid cystic carcinoma, pleomorphic adenoma; fibrous dysplasia, rhinosinusal extension like esthesioneuroblastoma, Wegener's granulomatosis.[2]
- **Metastatic:** Metastases from lung, kidney, breast, prostate, thyroid, colorectal origin can reach to the craniofacial bones and cause proptosis.[3]
- **Congenital:** Can be seen in brachycephaly, glial heterotopias,[4] Raine syndrome, cat's eye syndrome[5] and many other cranio-facial abnormalities.
- **Vascular:** Carotico-cavernous fistula, cavernous sinus pathologies, aneurysms

Classification

- **Painless and painful:** Painless are optic nerve gliomas, painful masses are neurofibroma (may be painless also), post-traumatic, superior ophthalmic vein thrombosis, etc.
- Visual compromise or not
- **Age: Children:** Hemangiomas, dermoids, optic nerve gliomas, neurofibroma (plexiform), meningiomas (less than adults), lymphoproliferative diseases, sarcomas from cone muscles. **Adults:** Cavernomas, neurofibromas (commoner than children), meningiomas, dermoids, rest all
- Unilateral or bilateral

REFERENCES

1. Maheshwari R, Weis E. Thyroid associated orbitopathy. *Indian Journal of Ophthalmology*. 2012;60(2):87.
2. Molina-Socola FE, Galvan-Carrasco MP, Del Estad-Cabello A, Sanchez-Vicente JL, Contreras-Diaz M, Rueda-Rueda T. Orbital tumour due rhinosinusal extension in a patient with Wegener's granulomatosis. *Archivos de la Sociedad Espanola de Oftalmologia*. Sep 2016;91(9):442–445.
3. Samlali H, Bouchbika Z, Bennani Z, et al. [Brain metastasis from rectal adenocarcinoma: about a case and review of the literature]. *The Pan African Medical Journal*. 2017;26:58.

4. Bakhti S, Terkmani F, Tighilt N, Benmouma Y, Boumehdi N, Djennas M. Glial heterotopia of the orbit: a rare cause of proptosis. Child's Nerv Syst 2016;32: 2239–2241.

5. Mosallanejad A, Sayarifard F, Hosseinverdi S, Abbasi F, Shabni Mirzaee H, Rezaei N. Proptosis, Micrognathia, Low Set Ear and Chest Deformity in a Patient with Extra Marker Chromosome 22. *Acta medica Iranica.* Dec 2015;53(12):782–784.

10 | Facial Pain

Facial pain

History

- Age and sex
- Onset
- Duration
- Progression
- Duration of each episode
- Side
  - Left
  - Right
  - Bilateral
- Location
- Radiation
- Character
- Trigger points
- Sleep disturbance
- Severity
- Associated
  - Skin changes
  - Lacrimation
  - Rhinorrhoea
  - Visual blurring
  - Decreased facial sensations
  - Redness of eyes
  - Pain on jaw movements
  - Jaw deviation or chewing problem
  - Depression and anxiety
  - Seizures
- Aggravating factors
- Relieving factors
- Trauma
- Procedural history
- Past history
- Drug history

Examination

- Inspection
- MMSE
- Vth nerve examination
- Sensorimotor examination
- TM joint examination
- Oral examination
- Psychological examination

The differentials of facial pain can be many including trigeminal neuralgia (TN), glossopharyngeal neuralgia (GN), atypical facial pain, dental pain and temporomandibular pain. For the scope of this book, we will focus upon the neurologic causes.

History

- **Age and Sex:** Young females may suffer from hysteria, middle aged persons suffer from TN, and elderly suffer from herpes zoster or giant cell arteritis. Some disorders are more common in females, e.g. TN.
- **Onset:** Onset of each episode should be enquired. Pain of TN is generally sudden in onset. Neuropathic pain is insidious in onset; psychogenic pain is of variable onset.
- **Duration:** Dental pains, oral mucosal inflammatory pains, traumatic pains, are generally recent onset ones. TN and atypical facial pain may be of short or long duration. Pain due to oral tumours, salivary gland tumours is generally of long duration. TM joint pains are commonly recent ones precipitated by prolonged opening of mouth in dental procedures, or chronic in cases of intra-articular disc problems.
- **Progression:** Initially the pain of trigeminal neuralgia starts with a single root distribution. Progressively in years the pain distributes to other root distribution. Pain of TN increases gradually, while pain of atypical facial pain may remain constant with periods of intermittent aggravation, with overall gradual increase over time.
- **Duration of each episode:** Periods of each episode of TN last for seconds to hours. Atypical facial pain (AFP) may cause continuous relentless pain. Pain of dental and TM joint origin is continuous and relentless.
- **Side:** Patient may complain of unilateral facial pain (right or left), or bilateral. TN and GN are generally unilateral. Atypical facial pain may be unilateral or bilateral. Burning mouth syndrome is bilateral.
- **Location:** Site of TN may be in one of the trigeminal root distribution, with most common in mandibular division, and maxillary the second one. Angle of mandible is spared, and pain never goes below chin. Site of atypical odontalgia may be in tooth or gums with history of tooth extraction and having no identifiable dental cause. TM joint pains are

generally present over the TM joint. Site of atypical facial pain is poorly localized with predominance over maxilla. Salivary stone pain is localised to sub-mandibular area. GN localises to deep inside the ear, back of tongue, tonsils or neck.

- **Radiation:** TM joint pains may radiate to muscles of mastication, temple, around the ear, and neck. Pain of atypical facial pain may cross the midline or anatomic boundaries of trigeminal nerve distribution. AFP may radiate to involve the temple, or occipital region.
- **Character:** It is very important as far as TN is concerned. Trigeminal neuralgia's character is lancinating and electric shock like. Pain of atypical facial pain may be of burning, nagging, diffuse or prickling in nature.
- **Trigger points:** Patients of TN may complain of trigger points in any of the trigeminal distributions, and is most commonly of mandibular distribution. Patient may complain of triggering pain by innocuous things like shaving, brushing, wind, etc. They may come without shaving of a part of a face.
- **Severity:** TN pain is generally severe incapacitating the patient for a short time period, while pain of AFP is of variable intensity.
- **Sleep disturbance:** TN generally do not disturb a patient's sleep, neither AFP does.
- **Associated symptoms:** Skin changes like shingles should be looked for as it may occur in herpes zoster. Lacrimation, visual blurring and rhinorrhoea may occur with ophthalmitis. Decreased facial sensations and redness of eyes, jaw deviation or chewing problems suggest involvement of trigeminal roots, although neurological deficits are generally absent in a case of TN. Trigeminal neurological deficits raise the possibility of schwannoma. Pain on jaw movements may suggest pathology of TM joint. Depression and anxiety are associated with AFP. AFP may also be associated with dysmenorrhoea, back pain, irritable bowel syndrome. Seizures may occur due to an epidermoid tumour lying in CP angle region.
- **Aggravating factors:** Patients with classical TN may complain of pain initiation with slightest touch over the face, some even cover the face to avoid wind touching their face.

Dental pains may aggravate or gets initiated with drinking hot or cold food or liquids. GN pain may get triggered with swallowing, coughing or touching the ear.

- **Relieving factors** TN patients get relieved with carbamazepine, but the effect of this drug sublimes because of its autoinduction property.
- **Trauma:** Trauma history should be taken always, as it may result in soft tissue, or bony injuries which may cause pain. It may also be associated with post-traumatic trigeminal neuropathy.
- **Procedural history:** Many patients come with a history of tooth extraction without any relief. Some come with history of neurectomy done in the past with temporary relief.
- **Past history:** History of stroke should be enquired.
- **Drug history**: patient may give history of taking carbamazepine, and if it gave benefit, then it is a pointer towards classic TN.

Examination

- **Inspection:** One should look for colour changes, shingles, corneal ulcers, unshaved faces, covered faces, extracted teeth, signs of trauma, neurectomy scars, jaw deviation.
- **MMSE:** To screen for disorders of cortical functions and dementia.
- **Vth nerve examination:** In a classical TN, it is essentially normal. Motor deficits suggest trigeminal schwannoma as the first likely diagnosis.

It includes motor and sensory examination.[2]

Motor examination: Inspect for hallowing of the temple region in temporalis atrophy, flattening of the jowl in atrophy of masseter muscle.

Place your fingers on the anterior border of masseters bilaterally. Ask the patient to clinch the teeth and feel the contraction of the muscle. Another way to test masseters is to place a wooden tongue blade in between molars, ask the patient to clinch the teeth, and see the markings. Unilateral weakness will cause absence or weak feeling of contraction in palpation, and markings on tongue blade.

Action of lateral pterygoid is tested by asking the patient to protrude the jaw. Unilateral weakness will result in ipsilateral directed deviation of jaw. It can also be tested by asking the patient to move the jaw side to side and see for weakness. Unilateral weakness will result in inability to move the jaw contralaterally.

It is difficult to examine the other muscles.

Sensory examination: Test for pain, touch, temperature in the same manner as examined elsewhere. Instead of asking whether they feel different, ask the patient whether the sensations are same. They should be checked in each branch distribution, as well as, peri-oral to posterior face in order to exclude onion-skin pattern of loss. Vibration testing can be used to test whether the loss is organic or non-organic. Place the tunic fork (128 Hz) over the frontal and mandible. Since these are single bones, any splitting of the response between the two sides suggests non-organic lesion.

Reflexes: Corneal reflex: It is elicited by lightly touching the upper cornea by a wet wisp of cotton. The response is sudden closure of bilateral lids. The afferent response is ipsilateral ophthalmic division and efferent response is facial nerve. If a stimulus is given in right eye with ipsilateral Vth nerve palsy there will be no response in any eye. But with a contralateral stimulus there will be normal consensual reflex in it.

Jaw jerk: Keep the patient's mouth open midway. Place your finger or thumb over the middle of chin and tap it with a hammer. The response is an upward jerk of the mandible. Normally it is minimal or absent. Its main use is to see hyperreflexia in a patient.

- **Sensorimotor examination:** To rule out lesion of the brainstem.
- **Temporomandibular joint:** Look for pain on movement, crepitus, tenderness.
- **Oral examination:** One should look for signs of inflammation over mucosa, dental caries, extracted tooth, excessive wear facets indicating bruxism.
- **Psychological** examination for atypical facial pain.

Investigations: Generally, the diagnosis is made by history and clinical examination. TN is especially a clinical diagnosis. However, investigations should be done to rule out other structural lesions particularly intracranial.

1. **MRI brain:** It is done primarily to exclude the intracranial masses, such as a trigeminal schwannoma or an epidermoid. MRI FIESTA sequence with thin slices can show a vascular loop compressing the Vth nerve exit zone. MRI may also show white matter plaques in multiple sclerosis.
2. **Panoramic X-ray of face**: It is done when a dental pathology is suspected.
3. **USG of salivary glands**: It is done when salivary lesions are suspected.
4. **ESR and CRP:** These are elevated in giant cell arteritis. ESR is generally >50.

Differential Diagnosis[3]

1. **Dental pain:** It is the commonest cause. Patient may localise the pain to a particular tooth. Hot or cold food/beverage will precipitate the pain. Causes may be dental caries, periodontal disorders, pericoronitis, etc.
2. **Trigeminal neuralgia:** Classically there is episodic pain of minutes to hours, has trigger points, involving distribution of one or more branches of trigeminal nerve. Most common is the mandibular division, and after it comes the maxillary division. There may be refractory period between the attacks. Character is sharp, shooting, electric-shock like, lancinating. It is severe and is triggered by light touch, washing, brushing, eating, or sometimes the wind. Causes may be vascular loop compression of root exit zone of trigeminal nerve (commonest is superior cerebellar artery, others are AICA, venous loops), arachnoid bands.
3. **Trigeminal post-herpetic neuralgia:** The symptoms are similar to the other neuralgias. Signs of herpes like shingles may be visible both extra and intraoral. It is a continuous pain of moderate (at times severe) severity with a character of burning, tingling or itch.
4. **Post-traumatic trigeminal neuralgia/trigeminal neuropathic pain:** It is caused by trauma to the face, dental procedures like tooth extraction. It appears with 3–6 months of trauma at its site and is continuous. The character is burning, tingling or sharp. It can be very severe, and may have triggering factors of touch, thermal or mechanical.

5. **Glossopharyngeal neuralgia:** The pain occurs unilaterally over deep in the ear, back of tongue, tonsils, or neck. Episodic duration is seconds to minutes. Character is similar to TN in being sharp, shooting, electric-shock like. It is aggravated by swallowing, coughing, or touching the ear.

6. **Short unilateral neuralgiform pain with autonomic features (SUNA/SUNCT) conjunctival injection and tearing:** It is a unilateral pain occurring mainly in first and second division of trigeminal nerve. Episodic duration is seconds to minutes, and is accompanied by autonomic features like tearing, red eye, rhinorrhoea, etc. The character is sharp or stabbing. The severity is moderate to severe.

7. **Giant cell arteritis:** Typical patient is an elderly female (male to female 1:2) with episodic pain over temple region. Pain may be bilateral, continuous with often sudden onset. The character is dull aching, throbbing. It is generally mild to moderate but may be severe, precipitated by chewing. There is accompanied visual blurring, diplopia, fever, myalgia, etc. ESR and CRP are elevated.

8. **Post-stroke pain:** It occurs ipsilateral to stroke over whole face and periorbital region. It begins after a few months of stroke or may be delayed. Character is pricking, aching, or burning. Severity is mild to moderate, and is precipitated by touch.

9. **Multiple sclerosis:** Facial pain is known to be an early sign of MS. Around 2.4% of TN patients suffer from MS.[4] Episodic duration is seconds to minutes, or sometimes hours. Character is electric shock or burning sensation. It is due to damage of the trigeminal nerve by the MS plaques. MRI confirms the diagnosis.

10. **Intracranial masses:** Trigeminal schwannoma may lead to facial pain. Characteristic of this lesion is accompaniment of motor weakness of trigeminal nerve along with the pain. Epidermoid tumour also may cause facial pain, which may present with other irritative symptoms like seizures. These posterior fossa masses may present with mass effect with symptoms and signs of hydrocephalous.

11. **Temporomandibular joint disease:** It is the second most common cause of facial pain after the dental origin. The

commonest cause is the prolonged oral opening in dental treatments or trauma. Second type is chronic one due to intra-articular disc problems, which presents with pain on opening and closing of the jaw, clicking sounds, and locking. Subluxation of the joint may also be a pathology and results in deviation of the jaws on opening.

12. **Burning mouth syndrome:** It is a disorder of menopausal or post-menopausal women. Patient may complain of burning pain bilaterally in and around the tongue and oral mucosa. Examination is essentially normal. It is considered to be due to disorder of peripheral trigeminal nerve fibres or brainstem pathology.[5] Eating may aggravate or relieve the pain.

13. **Atypical facial pain:** It is idiopathic with non-anatomic distribution. It may be continuous or has episodes of hours or days. The character is dull-aching, or sometimes sharp. Severity is mild-moderate and precipitated by fatigue, stress, life events, bowel problems, etc.

Table 10.1: Classification scheme for trigeminal neuralgia (TN) and related facial pain syndromes[6]	
Type	*Description*
TN1	Classic or typical TN. Idiopathic, spontaneous facial pain which is predominantly episodic in nature
TN2	Atypical TN, or TN type 2. Idiopathic spontaneous facial pain which is predominantly constant in nature
TN3	Trigeminal neuropathic pain. Results from unintentional injury to the trigeminal nerve from trauma to surgery
TN4	Trigeminal deafferentation pain. Results from intentional injury to the nerve in an attempt to treat either TN or other related facial pain
TN5	Symptomatic TN. Results from multiple sclerosis
TN6	Postherpetic TN. Follows a cutaneous herpes zoster outbreak in the trigeminal distribution
TN7	"Atypical facial pain". Facial pain secondary to a somatoform pain disorder, requiring psychological testing to confirm

REFERENCES

1. Bayer DB, Stenger TG. Trigeminal neuralgia: an overview. *Oral Surg Oral Med Oral Pathol.* 1979;48(5):393–399.

2. Campbell WW. *DeJong's THE Neurologic Examination.* Seventh. Philadelphia: Lippincott Williams & Wilkins; 2013.

3. Zakrzewska JM. Differential diagnosis of facial pain and guidelines for management. *Br J Anaesth.* 2013;111(1):95–104. doi:10.1093/bja/aet125.

4. Jensen TS, Rasmussen P, Reske Nielsen E. Association of trigeminal neuralgia with multiple sclerosis: clinical and pathological features. *Acta Neurol Scand.* 1982;65(3):182–189.

5. Forssell H, Jaaskelainen S, Tenovuo O, Hinkka S. Sensory dysfunction in burning mouth syndrome. *Pain.* 2002;99(1–2):41–47.

6. Eller JL, Raslan AM, Burchiel KJ. Trigeminal neuralgia: definition and classification. *Neurosurg Focus.* 2005;18(5):E3.

Facial Asymmetry

Facial asymmetry

History
- Onset
- Duration
- Progression
- Side
- Corneal ulceration
- Lacrimation
- Taste
- Hearing difficulty
- Components of facial muscle paralysis
  - Flattening of forehead
  - Decreased or absent blinking
  - Food sticking in the bucco-gingival sulcus
  - Biting of cheek
  - Saliva drooling from the angle of mouth
- Asymmetry more noticed when involuntarily/ voluntarily smiling
- Facial muscle fasciculations
- Hemifacial spasm
- Facial myokymia
- Headache
- Other neurological deficits
- Posterior auricular pain
- Vesicles
- Fever
- Trauma
- Past history
  - Viral infection
  - Stroke
  - DM
- Drug history

Examination
- General examination
- Facial expression
- Facial nerve examination
  - Motor
  - Sensory taste
  - Secretory: Schirmer's test
- Higher mental functions
- Other cranial nerves
- Motor and sensory examination
- Skin over the ear and parotid

Most of the facial asymmetry cases are due to facial nerve palsies. In it, we have to make out whether it is lower motor neuron (LMN) or upper motor neuron (UMN) type. With history, examination and investigations, we have to then localise and make out the differential diagnosis. Other causes of facial asymmetry may be due to disorders of facial expressions, like extrapyramidal disorders.

History

- **Onset:** Sudden onset facial palsy, often within 72 hours, may occur with Bell's palsy, stroke, diabetes mellitus, herpes infection, parotid abscesses, GBS, and trauma. Insidious onset may occur with neoplasms. If congenital facial palsy is found, then history of forceps delivery should be sought.
- **Duration:** In long duration facial palsy (more than 3 weeks), neoplastic pathology should be suspected.
- **Progression:** Bell's palsy does not progress, and recovers with the help of corticosteroids. Facial palsy due to stroke does not progress after initial event, and may or may not be permanent. A neoplastic cause has a progressive course.
- **Side:** Bell's palsy, neoplasms, stroke may present with unilateral facial palsy. Bilateral facial palsy may occur with Bell's palsy (1% only), bilateral CP angle neurofibromas, Lyme disease, Guillain-Barré syndrome, sarcoidosis, and meningitis.
- **Corneal ulceration:** Decreased eye blinking, and inability to close the eyes may result in exposure keratitis and corneal ulceration. Closure of the eyelids is a function of levator orbicularis occuli muscle supplied by frontal and zygomatic branch of facial nerve. History of open eyes during sleep may be taken from the relatives to delineate subtle facial palsy.
- **Lacrimation:** History of decreased lacrimation should be asked. The lacrimal glands are located in supero-lateral quadrant of the eyes. The parasympathetic supply (which is responsible for lacrimation) comes from lacrimatory nucleus of facial nerve in the pons. Complete pathway is shown in Fig. 11.1.
- **Taste:** Decreased taste should be enquired about, though it may be reported by only 1/3rd of patients due to preserved sense of taste in opposite half of tongue.[1] Taste sensations are carried to nucleus tractus solitarius through the fibres of

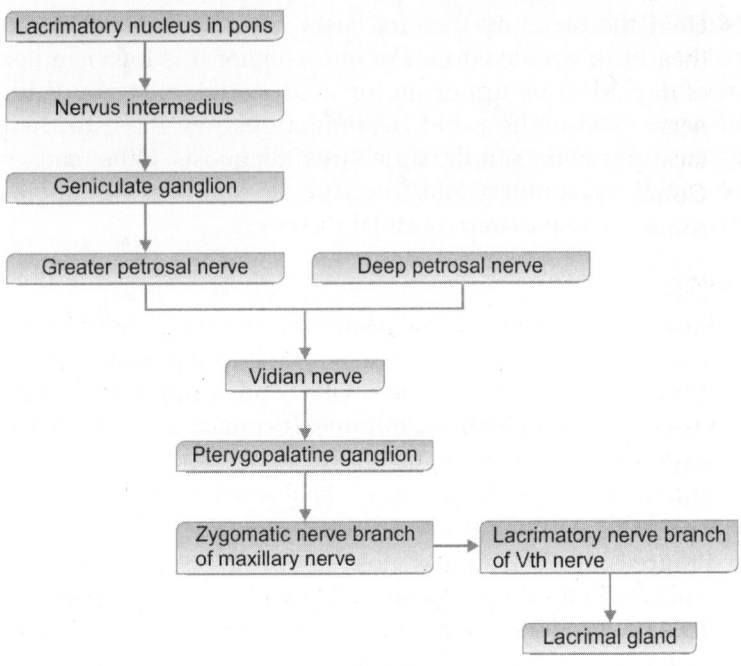

Fig. 11.1: Lacrimatory pathway

VIIth nerve, from where these are carried to gustatory cortex, as shown in Fig. 11.2.

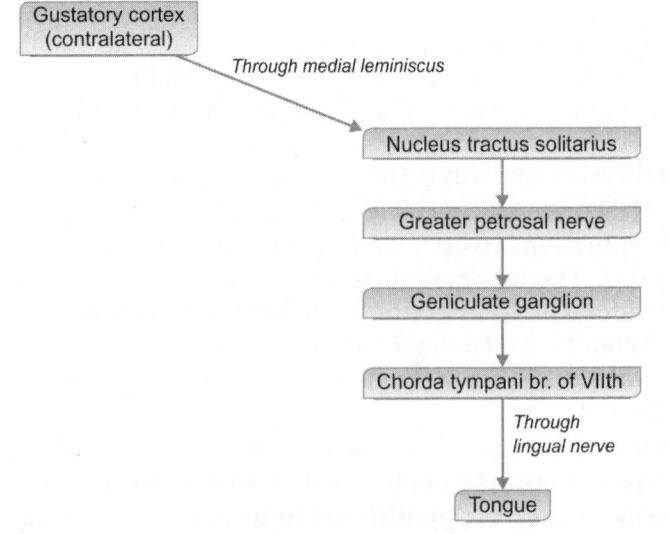

Fig. 11.2: Taste pathway

- **Hearing difficulty:** Patients may complain of hyperacusis (hearing normal sounds louder). This is because of the loss of dampening effect of stapedius muscle supplied by facial nerve. Decreased hearing, tinnitus and vertigo along with facial asymmetry may occur due to lesion in CPA region.

- **Components of facial muscles paralysis:** Patient may complain of loss of forehead wrinkling, inability to close the eyes, food sticking in the bucco-gingival sulcus, biting of cheek, food and saliva drooling from angle of mouth, change in speech (due to difficulty in pronouncing labials). If forehead wrinkling and eye blinking is normal, then the facial palsy is upper motor neuron type.

- **Asymmetry more** noticed when involuntarily or voluntarily smiles: Asymmetry on voluntary smiling means lesion is in the pyramidal pathway (including frontal lobe and corticobulbar pathway), while asymmetry on involuntary smiling is because of lesion of the extrapyramidal pathway (including thalamic and striatocapsular pathways).

- **Facial muscle fasciculations:** These may occur in LMN type lesion

- **Hemifacial spasm:** Patients may complain of involuntary tonic contraction of one side of face. These may occur even in sleep. The spasm begins in the orbicularis oculi, and progresses downwards.

- **Facial myokymia:** It is a unilateral continuous quivering of facial muscles.

- **Headache:** Its association with facial palsy may point towards a cranial neoplastic lesion.

- **Other neurological deficits:** One should look for other cranial nerve palsies which can point towards brainstem lesion or its compression like hearing loss, tinnitus, vertigo (CPA lesion), contralateral hemiplegia and numbness (lobar or brainstem involvement).

- **Posterior auricular pain:** Patients may experience posterior auricular pain in Bell's palsy,[1] which precedes paralysis by 2–3 days. Pain may also occur in temporal bone fractures due to trauma.

- **Vesicles:** Herpes zoster infection may also be associated with facial palsy, which presents with blisters over and around the auricle (also called Ramsay Hunt syndrome).

- **Fever:** Parotid abscess may be associated with fever. Flu-like symptoms, joint pain, rash may occur due to Lyme disease.
- **Trauma:** Temporal bone fractures may be associated with LMN type facial nerve palsy, while contusions proximal to pons may lead to UMN type facial palsy.
- **Past history:** History should be taken about viral infections, stroke, diabetes mellitus which can be associated with facial palsy.

Examination

- **General examination:** One should see the orientation of the patient, apart from the other general examination variables.
- **Facial expressions:** Patients of neurologic illnesses may have abnormal facies. In Parkinsonism patient presents with mask-like facies; progressive supranuclear palsy patient has facial dystonia or omega sign (knitting of the eyebrows with wide palpebral fissure).
- **Facial nerve examination:**[2] This is the most important part of the examination. This should be examined for motor and sensory functions.
  - **Motor functions:**
  - Look for the symmetry of the face. It is generally symmetrical in young age, and may have slight asymmetry due to the ageing character lines, that is non-pathological.
  - One should look for atrophy and fasciculations, which may point towards LMN type of lesion.
  - Ask patient to wrinkle his forehead. See for its symmetry. Absence of it suggests LMN type facial nerve palsy.
  - Observe spontaneous eye blinking for frequency and symmetry—these are affected in LMN lesions. Unilateral palpebral fissure may be widened due to loss of tone of orbicularis oculi.
  - Ask patient to close his eyes. Normally a patient can close his eyes. On forceful closure, one cannot open the eyes by gentle lifting of the eyelids by little finger. This exercise can also be compared on both sides to see for subtle weakness. Its a test for orbicularis oculi. Another sign for subtle lesions is to ask the patient to tightly close the eyes and feel the vibrations by placing your thumb over them. Normally there is fine vibrations, which are attenuated in the palsy—known as Bergara-Wartenberg sign.

– Observe nasolabial folds—loss of them may be seen both in UMN and LMN lesions, but more pronounced in LMN lesions.

– Ask the patient to voluntary pull the angles of the mouth down in an exaggerated manner. This is the test for platysma.

– Ask him to whistle and fill air in the mouth, and hold it while you try to press the cheeks. Its a test for orbicularis oris.

– Observe spontaneous movements of face as patient talks, smiles or frowns. Weakness only in these involuntary acts occur in thalamic or striatocapsular lesions.

– Patients of hemifacial spasm may show spontaneous contraction of the face.

– **Sensory functions:** Examine for **taste.** For this take solutions of sugar for sweet taste, table salt for salty taste, citric acid for sour taste, and quinine for bitter taste. Make cards for all these sensations, show and explain them to the patient. Thereafter, explain the procedure and hold the patient's tongue with a gauge piece and apply these solutions, sequentially (bitter last), first on the abnormal side and then on the normal side. Ask the patient to point towards the relevant card if he feels the sensation (do not leave his tongue—otherwise the solution will spread bilaterally). Patient should rinse his mouth between changing from one solution to other.

The solution should be applied at the junction of the anterior and middle thirds of tongue. Most patients can identify within 10 seconds of application.

• **Secretory functions:** The **Schirmer's test** is used for this. Special filter paper (Whatmann No. 41) is used and is placed in the inferior conjunctival sac for 5 minutes. It is done in two manners, and for both eyes.

Schirmer's test I (without anaesthesia): After wiping out the eyes, fold the Schirmer paper at 5 mm mark and place it in the inferior conjunctival sac for 5 min. See for how much it becomes wet after this time. If the paper becomes completely wet before 5 min, then the test can be terminated prematurely. The folded 5 mm is subtracted. Normal values are 15–30 mm. Less than 10 mm suggests moderate dry eye, and less than 5 mm suggests severe dry eye.

Schirmer's test II (With topical anaesthesia): This is done to see for keratitis sicca.

Table 11.1: House Brackmann grading for motor functions[4]	
Grade I:	Normal
Grade II:	Mild dysfunction. Slight weakness observed on close inspection. Forehead: Moderate to good function; Eye: Complete closure with minimal effort; Mouth: Slight asymmetry.
Grade III:	Moderate dysfunction. Obvious but not disfiguring. Forehead: Slight to moderate movement; Eye: Complete eye closure with effort; Mouth: Slight weakness with maximum effort.
Grade IV:	Moderately severe dysfunction. Obvious weakness and/or disfiguring. Forehead: None; Eye: Incomplete closure; Mouth: Asymmetry even with maximum effort.
Grade V:	Severe dysfunction. Barely perceptible motion. Forehead: None, Eye: Incomplete closure; Mouth: Slight movement
Grade VI:	Total paralysis

- **Higher mental functions:** To rule out lobar lesions, dementia, stroke.
- **Other cranial nerves:** Vth nerve (*see* chapter: Facial Sensory Loss), VIth nerve (*see* chapter: Diplopia), VIIIth (*see* chapter: Hearing Loss) and lower cranial nerves (*see* chapter: Dysphagia), examination should be done to rule out brainstem and cerebellopontine angle lesions.
- **Motor and sensory examination:** It should be done to rule out stroke, brainstem lesions (brainstem masses, tuberculomas, cavernomas, etc.), or compression of the long tract signs by a lesion near their vicinity (vestibular schwannoma, meningioma, epidermoid, arachnoid cyst, etc.)
- **Skin around the ear and parotid:** To look for vesicles. Parotid tumours may lead to LMN type facial palsy. Skin over the parotid gland should be examined as it may be red and hot in case of parotid abscess.

Investigations

1. **CT Head:** It is done to see for intracranial pathologies. Apart from the lesions, it may also show bone erosions like petrous tip erosion in CPA tumours, widening of IAC in vestibular schwannoma, hydrocephalous, etc.

2. **MRI Head:** In tumours it may help out in making diagnosis and plan out the surgery. In multiple sclerosis it will show the characteristic plaques.
3. **Lumbar puncture with CSF analysis:** In suspected meningitis, GBS.
4. **IgG and IgM** for *Borrelia burgdorferi* and herpes simplex: If Lyme disease or herpes infection is suspected.
5. **Blood sugar and HbA1c:** To rule out diabetes mellitus.

Localisation

- **UMN facial palsy:** Involvement of only lower face, with sparing of forehead and eye blinking, localises the lesion as UMN type. This is because upper face has bilateral supply. UMN type facial palsy can be volitional or emotional. It is generally associated with hemiplegia.

  Weakness appearing only on voluntary smiling is known as volitional and occurs in contralateral precentral gyrus lesions or corticobulbar tract. Weakness appearing in spontaneous smiling, and absent in voluntary smiling, occurs due to lesions in thalamus and striatocapsular lesions.

- **LMN facial palsy:** Affection of all functions including lacrimation localises the lesion proximal to geniculate ganglion (as greater petrosal nerve arises from it). It may include lesions at nuclear, cisternal segment, and labyrinthine segment. Nuclear lesions will have fasciculations present. Association with tinnitus, deafness, vertigo, facial pain localises the lesion in the cisternal segment, e.g. vestibular schwannoma.

  Sparing of lacrimation and involvement of taste localise the lesion to be proximal to chorda tympani nerve, i.e. in the facial canal.

  Preservation of lacrimation and taste, and involvement of motor functions localise the lesion to be after chorda tympani, i.e. in the lower part of facial canal or in parotid gland. Diabetes in 83% of cases causes ischemic neuropathy in the facial canal.[3]

- **Hemifacial spasm:** It generally occurs due to compression of facial nerve rootlets due to an aberrant vessel, most commonly AICA. Other causes may be tumour, demyelination or rarely due to Bell's palsy.

- **Facial myokymia:** It occurs in multiple sclerosis, pontine tumours, compressive lesions in CPA region (tumours), GBS, basilar invagination.

Differential Diagnosis

- **Bell's palsy:** Idiopathic. LMN type facial palsy occurs. It is a diagnosis of exclusion.

- **Stroke:** Generally upper motor neuron type facial palsy occurs. With facial palsy contralateral hemiplegia and numbness may also be associated.

- **Neoplasms:** Both UMN and LMN type facial palsy may occur depending upon the location of the mass. Neoplasms causing UMN type facial palsy include tumours near lower part of motor cortex, tumours displacing capsular fibres, thalamic tumours, and ganglionic lesions. For LMN type palsy tumours may occur in brainstem (gliomas, tuberculomas, cavernomas, etc.), CP angle region (vestibular schwannoma, meningioma, epidermoids, arachnoid cyst, lower cranial nerve schwannoma facial nerve schwannoma, lipomas, etc).

- **Lyme disease:** Its a bacterial infection (*Borrelia burgdorferi*) caused by tick. Along with the facial palsy, a pink or red rash may occur around the bitten area. Flu-like symptoms, joint pain or tiredness may also occur.

- **Parotid tumour/abscess:** Both benign and malignant parotid tumours, and parotid abscess may cause facial nerve palsy.

- **Ramsay Hunt syndrome:** Facial nerve palsy with blisters over ear.

- **Neurosarcoidosis:** VIIth is the most commonly involved nerve in this entity. It may be unilateral or bilateral. Other features include diffuse sensorimotor neuropathy, distal to proximal slowly progressive weakness, multifocal neuropathy mimicking mono-neuritis multiplex, localised or generalised muscle weakness or soreness. An acute generalised motor neuropathy like GBS may also occur.

- **Fascioscapulohumeral dystrophy:** There is weakness of facial and eye muscles in this genetic disease. Other features are weakness of shoulders, lower leg, hips, abdominal muscles.

Other rare differential diagnosis

- Aneurysms of vertebral, basilar or carotid arteries
- Acute or chronic otitis media
- Amyloidosis
- Autoimmune syndromes
- Botulism
- Cholesteatoma of middle ear
- Glomus tumours
- Guillain-Barré-syndrome

REFERENCES

1. Peitersen E. The natural history of Bell's palsy. *Am J Otol.* 1982;4(2):107–111.

2. Campbell WW. *DeJong's THE Neurologic Examination.* Seventh. Philadelphia: Lippincott Williams & Wilkins; 2013.

3. Pecket P, Schattner A. Concurrent Bell's palsy and diabetes mellitus: a diabetic mononeuropathy? *J Neurol Neurosurg Psychiatry.* 1982;45(7):652–655.

4. House JW, Brackmann DE. Facial nerve grading system. Otolaryngol Head Neck Surg 1985;93: 146–147.

12 Taste

CHAPTER

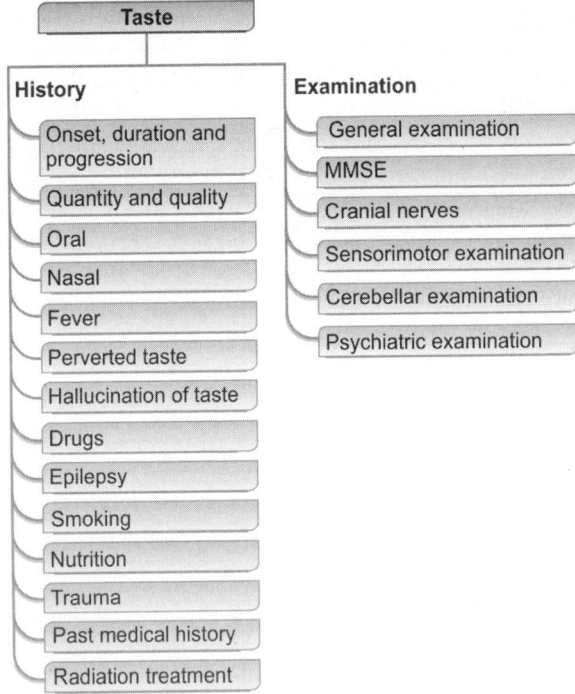

Taste

History	Examination
Onset, duration and progression	General examination
Quantity and quality	MMSE
Oral	Cranial nerves
Nasal	Sensorimotor examination
Fever	Cerebellar examination
Perverted taste	Psychiatric examination
Hallucination of taste	
Drugs	
Epilepsy	
Smoking	
Nutrition	
Trauma	
Past medical history	
Radiation treatment	

Introduction: Taste and smell are interrelated. While chewing, odours of different foods are released, which reach the olfactory system located in posterior nasal pharynx. Tongue perceives only sweet, sour, salty, bitter, and metallic taste.

Taste impairment can be physiologic or pathologic. There are natural variation in taste perception as detected by taste studies using phenylthiocarbamide (PTC). In old age, taste perception decreases. Pathologic causes can be due to many reasons, which we will discuss. Pathway is shown in Fig. 11.2, chapter: Facial Asymmetry.

History

- **Onset, duration and progression:** Causes of acute onset ones are vascular lesion, head injury, Bell's palsy, URTI, psychiatric condition. Subacute ones are meningitis, encephalitis. Chronic ones are neoplasm (CP angle, brainstem, etc.), motor neuron disease, etc.
- **Quantity and quality of taste:** Ask, whether the taste is normal or abnormal; if abnormal then it is more or less. Absent taste sensation is known as aguesia, while decreased is known as hyposguesia. Dysguesia includes complaints like metallic taste, or a permanent bitter, sour, salty, or rarely sweet. Qualitative taste abnormalities is known as paraguesia or phantoguesia.
- **Oral:** Oral lesions like ulcerations, caries, etc. may affect the taste.
- **Nasal:** Catarrh, adenoids, etc. may affect the taste.
- **Fever:** In fever the taste gets altered.
- **Perverted taste:** It is seen in pregnancy, hysteria, glossitis, epilepsy.
- **Hallucination of taste:** It is seen in mental disease, head injury, etc.
- **Drugs:** They may alter the taste by either increasing or decreasing salivation, or getting directly secreted in saliva.
- **Epilepsy:** Jacksonian epilepsy, uncinate fits may localise the lesion to uncus.
- **Smoking:** Heavy smokers of 20 or more/ day impair taste in approximately 20% of the population.[1]
- **Nutrition:** Chronic malnutrition, particularly vitamin B_3, B_{12}, and zinc may lead to impaired taste.
- **Trauma:** This should be enquired to rule out facial nerve pathway damage.
- **Past medical history:** Diabetes, sarcoidosis, multiple sclerosis, chronic renal failure, liver cirrhosis, cancers, etc. can lead to impaired taste.
- **Radiation treatment:** Head and neck radiation may alter taste sensations.

Examination

- **General examination:** Examine for nutritional status, fever. Look for rhinitis, signs of head trauma, oral or perioral skin lesions (viral like herpes), white plaque on tongue (candidiasis, HIV, leukoplakia). Ears should be examined for discharges; eyes should be examined for dryness. Look for thyroid enlargement, parotid and submandibular gland enlargement and previous neck surgeries.
- **Neurological examination**
  - ➢ **MMSE:** It should be done to look for Alzheimer's disease. If found decreased, one should do detailed higher mental function examination.
  - ➢ **Cranial nerves:** Examine for facial nerve. (Explained in Facial Asymmetry chapter, only the taste examination being explained here.)
    - – Examine for **taste.** For this take solutions of sugar for sweet taste, table salt for salty taste, citric acid for sour taste, and quinine for bitter taste. Make cards for all these sensations, show and explain them to the patient. Thereafter, explain the procedure and hold the patient's tongue with a gauge piece and apply these solutions (either with a dropper or taste strip), sequentially (with bitter last): citric acid (sour), glucose (sweet), sodium chloride (salty), and quinine (last), first on the abnormal side and then on the normal side. The solution should be applied at the junction of the anterior and middle thirds of tongue. Most patients can identify within 10 seconds of application.

      Ask the patient to point towards the relevant card if he feels the sensation (do not leave his tongue—otherwise the solution will spread bilaterally). Patient should rinse his mouth between changing from one solution to other.

      If not recognised at all, means absence of taste. While if recognised in a lesser degree, means impairment of taste, which is seen in oral, nasal lesions and fever.

      Misinterpretation of taste is seen in perverted taste, and experiencing taste without any external stimulation means hallucination.
  - ➢ **Other cranial nerves:** Examination of other nerves should also be done to help in localisation of the lesion. Testing

of olfaction should always be done as this is often lost with taste.

> **Sensorimotor examination:** Hemiplegia with 5th and 7th involvement localises to pontine region.

> **Cerebellar examination:** Should be done for localisation purposes. Its impairment suggests ipsilateral posterior fossa lesion.

- **Psychiatric examination:** If psychiatric problem is suspected, it can be done.

Investigations

- **Laboratory tests:** These can be ordered to rule out medical diseases on the basis of history and physical examination. These are blood sugar levels, renal, liver, and thyroid function tests, HIV status, B_{12}, serum ferritin, folate, and zinc levels, etc.

- **Imaging: CT head:** It is a good investigation to rule out temporal bone fractures, middle ear pathologies, skull base lesions, intracranial bleed, contusion or infarct.

- **MRI:** It can be done when an intracranial pathology like multiple sclerosis, tumours, abscesses, etc. are suspected. It can also be used in skull base lesions. Gadolinium can be used whenever needed.

- **EEG:** When epilepsy is thought in the differential diagnosis, this is a valuable tool. It also shows slowing of background activity in Alzheimer's disease.

Differential Diagnosis

- **Post-infectious:** URTIs may impair taste sensations transiently and in reversible manner. Otitis media may affect chorda tympani fibres.[2]

- **Post-surgical:** Dental surgery involving wisdom tooth extraction, local anaesthesia instillation at that region may injure lingual nerve and cause taste disturbance.[3] Tonsillectomy may cause transient (up to 1/3rd) or permanent (up to 1%) taste disturbances.[4] Dysguesia is more common than complete absence. Middle ear surgery,[5] maxillofacial and cranial surgery[6] may also cause taste disturbances.

- **Neurological conditions: Bell's palsy and Ramsay Hunt syndrome** affect taste sensations in about 30% cases by impairing facial nerve function. This is generally transient.[7]

Stroke in region of taste pathways may impair this sensation in about 20% of cases, but have good prognosis.[8] **Neurodegenerative diseases** like Alzheimer's dementia, Parkinson's disease, ALS, multiple sclerosis, may affect taste. Dysguesia may be the first sign of ALS.[9] Dysguesia may manifest as seizure activity in **Epilepsy**.[3] Myasthenia gravis and Guillain-Barré syndrome may also cause dysguesia.

- **Head trauma:** Temporal bone fractures, intracranial contusions, etc. may impair taste by affecting taste pathways.
- **Metabolic causes:** Diabetes, thyroid disease, liver cirrhosis, chronic renal failure, Sjögren's syndrome, may manifest with taste disturbances.
- **Drugs:** These are antibiotics (metronidazole, azithromycin, ciprofloxacin, tetracycline), anticonvulsants (carbamazepine, phenytoin), lipid lowering agents (lovastatin, pravastatin), antidepressants (amitryptiline, nortryptiline, imipramine, clomipramine), antimanic (lithium), antihistamines (chlorpheniramine, loratidine, pseudoephedrine), antihypertensives and cardiac drugs (acetazolamide, nitroglycerin, spironolactone), anti-inflammatory (colchicine, penicillamine), antineoplastic (methotrexate, cisplatin, vincristine), anti-parkinsonian (levodopa), muscle relaxants (baclofen, dantrolene).
- **Neoplastic:** Tumours of **cerebellopontine angle** often cause taste disturbances by impairing facial nerve. **Paraneoplastic** causes include small cancer of lung, and thymoma.
- **Nutritional deficiencies:** These include vitamin B_{12} (may accompany glossitis), iron, folate and zinc deficiency.

REFERENCES

1. Vennemann MM, Hummel T, Berger K. The association between smoking and smell and taste impairment in the general population. *Journal of neurology.* 2008;255(8):1121–1126.
2. Landis BN, Beutner D, Frasnelli J, Hüttenbrink KB, Hummel T. Gustatory function in chronic inflammatory middle ear diseases. *The Laryngoscope.* 2005;115(6):1124–1127.
3. Landis BN, Lacroix JS. Taste disorders. *B-ent.* 2009;5 Suppl 13:123–128.
4. Heiser C, Landis BN, Giger R, et al. Taste disorders after tonsillectomy: a long-term follow-up. *Laryngoscope.* Jun 2012;122(6): 1265–1266.

5. Kveton JF, Bartoshuk LM. The effect of unilateral chorda tympani damage on taste. *Laryngoscope.* Jan 1994;104(1 Pt 1):25–29.

6. Cullen MM, Leopold DA. Disorders of smell and taste. *The Medical clinics of North America.* Jan 1999;83(1):57–74.

7. Heckmann JG, Heckmann SM, Lang CJ, Hummel T. Neurological aspects of taste disorders. *Archives of neurology.* May 2003;60(5):667–671.

8. Heckmann JG, Stossel C, Lang CJ, Neundorfer B, Tomandl B, Hummel T. Taste disorders in acute stroke: a prospective observational study on taste disorders in 102 stroke patients. *Stroke.* Aug 2005;36(8):1690–1694.

9. Petzold GC, Einhaupl KM, Valdueza JM. Persistent bitter taste as an initial symptom of amyotrophic lateral sclerosis. *Journal of neurology, neurosurgery, and psychiatry.* May 2003;74(5):687–688.

Hearing Loss

```
                    ┌─────────────────┐
                    │  Hearing loss   │
                    └────────┬────────┘
          ┌──────────────────┴──────────────────┐
  History                                   Examination
  ┌───────────────────┐               ┌───────────────────┐
  │ Which ear         │               │ MMSE              │
  ├───────────────────┤               ├───────────────────┤
  │ Age               │               │ Otoscopic         │
  ├───────────────────┤               ├───────────────────┤
  │ Onset, duration   │               │ Mastoid region    │
  │ and progression   │               ├───────────────────┤
  ├───────────────────┤               │ Cochlear          │
  │ How it was first  │               ├───────────────────┤
  │ noticed           │               │ Vestibular        │
  ├───────────────────┤               ├───────────────────┤
  │ Pitch             │               │ Other cranial     │
  ├───────────────────┤               │ nerves            │
  │ Hearing in        │               └───────────────────┘
  │ crowded places    │
  ├───────────────────┤
  │ Recruitment       │
  ├───────────────────┤
  │ Speech            │               ┌──────────────────┐
  │ discrimination    │               │ Tinnitus         │
  ├───────────────────┤               ├──────────────────┤
  │ Tone decay        │               │ Vertigo          │
  ├───────────────────┤               ├──────────────────┤
  │ Medications       │               │ Discharge        │
  ├───────────────────┤──────────────>├──────────────────┤
  │ Associated        │               │ Itching          │
  │ symptoms          │               ├──────────────────┤
  ├───────────────────┤               │ Earache          │
  │ Trauma            │               ├──────────────────┤
  ├───────────────────┤               │ Vesicular        │
  │ Exposure to loud  │               │ eruptions        │
  │ sounds            │               └──────────────────┘
  ├───────────────────┤
  │ Speech difficulty │
  ├───────────────────┤
  │ Aggravating/      │
  │ relieving factors │
  ├───────────────────┤
  │ Family history    │
  ├───────────────────┤
  │ Treatment taken   │
  │ and present status│
  └───────────────────┘
```

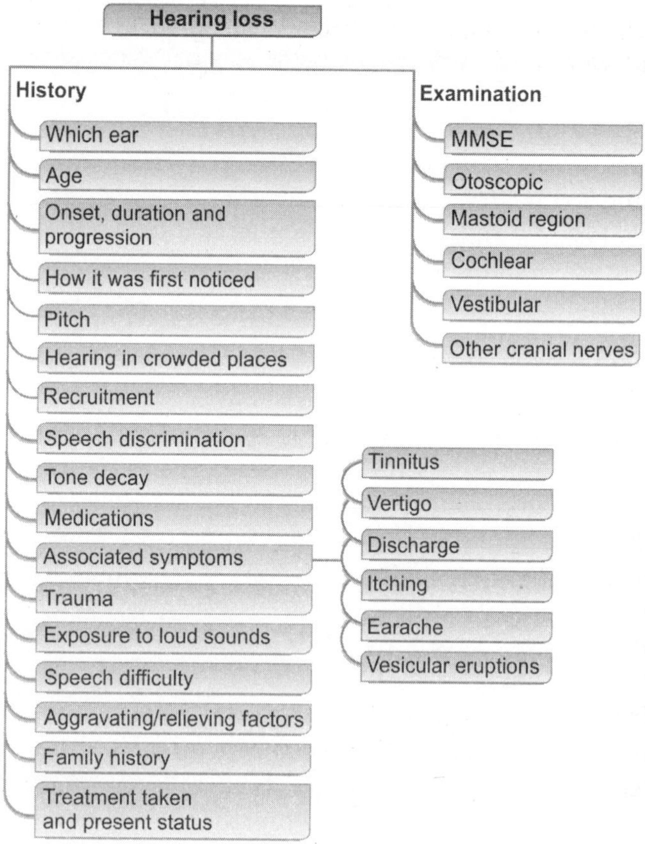

Generally, patients with hearing loss are initially seen by ENT surgeons, and on exclusion of the conductive hearing loss (CHL), they are referred to us. Primary aim of this chapter is to tailor our history and examination in a way to differentiate between CHL and sensorineural hearing loss (SNHL), and reach to the neurologic diagnosis.

History

- **Which ear:** Hearing loss in which ear: Right, or left, or both should be enquired.
- **Age:** Childhood causes of deafness are congenital ones, syphilis, birth injury, anoxia. Adult onset includes otosclerosis, MS, syphilis, hysteria. Old age ones are tumours, senile deafness.
- **Onset, duration and progression:** Sudden onset causes include intratumoral haemorrhage, vascular like internal auditory artery occlusion, viral infection, trauma. Subacute onset is seen in infections and inflammations. Chronic onset is seen in neoplasm, MS, Meniere's disease, etc.
- **How it was first noticed:** Difficulty in using cell phone, could not hear voice when called from involved side, not able to hear/appreciate vehicles/horn of vehicles passing from involved side, difficulty in normal conversation, noticed by family members, started to speak in louder voice noticed by family members.
- **High pitch or low pitch hearing loss:** High pitch (sensorineural hearing loss). High pitch includes feminine voice, cry of baby, ringtone, ringing of door bell, falling utensils, chirping of birds. Low pitch hearing loss (conductive hearing loss): Male voices, loud sounds (drums), watching television, sounds of freeze and washing machine.
- **Hearing in crowded places:** There is high frequency hearing loss in SNHL. Therefore, due to background noise which is low frequency patient may find out of proportion difficulty in conversing. This is absent in CHL.
- **Hearing improves in noisy surroundings:** This occurs in Paracusis Willisi (Otosclerosis: CHL)
- **Recruitment phenomenon (SNHL):** There is exaggerated response to perception of sound. A little increase in the intensity of sound will feel louder, distorted and discomforting. Patient may give history of sudden painful increase in sound on progressively increasing television volume. This is because due to loss of some hair cells, the neighbouring hair cells have to take up their function. These

recruited hair cells function for their frequencies as well as the damaged hair cells frequencies.

- **Speech discrimination:** Able or not able to make our content of speech (SNHL), due to loss of high frequency hearing.
- **Tone decay:** History taking of this point is difficult. Patient may give history of low intensity sounds decaying after sometime. In CHL, this decay is only for sounds 0–5 dB above their absolute hearing threshold, in cochlear it is for 10–25 dB, and for retrocochlear lesions it is for >25 dB above their thresholds.
- **Medications instilled in ear:** Streptomycin, antimalarials, pain killers, antihypertensive agents (diuretics), heavy metals, toxins may lead to hearing loss.
- **Associated symptoms:** Tinnitus, vertigo, ear discharge, itching, earache, vesicular eruptions should be asked.
- **Trauma** history to the ear/bleeding from the ear should be enquired.
- **Exposure to loud sounds:** Noise induced hearing loss is the most common because of damage to hair cells, though it may also be because of rupture of tympanic membrane, and damage to ossicles. It can be both temporary or permanent.[1]
- **Speech or language difficulty:** Hearing is critical to speech and language development. Therefore, these difficulties are often seen in association with hearing loss.
- **Aggravating/ relieving factors:** Waxing or waning hearing loss is seen in wax (conductive deafness), Meniere's disease (SNHL), and perilymphatic fistula (SNHL).
- **Family history of hearing loss:** This is relevant for neuro-fibromatosis 2.
- Treatment taken and effect, and present status.

Examination[2]

MMSE: It should be done to rule out dementias and cortical involvement. If found abnormal detailed higher mental function examination should be done.

Otoscopic: Using otoscope see for wax, blood, foreign bodies, or tympanic membrane perforation. Wax or cerumen impaction is the most common cause of hearing complaint,[3] and hearing loss can range from 5–40 dB depending upon the impaction degree.[4]

Mastoid region: Examine this region for swelling and tenderness. Bruising over this region is found in traumatic middle cranial fossa fractures resulting into path of blood along posterior auricular artery, known as Battle's sign.

Cochlear Nerve Examination

1. **Observe the patient** during conversations. Note his head turning, lip reading, or speaking with loud voice.

2. **Whispered voice test:** In a quiet room, whisper one by one in both ears of the patient, as much as possible, in equal amplitude and from equal distance. Any difference suggests abnormality. It is a good test for screening, though having theoretical limitations of nearly impossible feature of equal amplitude and distance.

3. **Finger rub test:** Rub your thumb and finger at equal amplitudes and from equal distance.

4. **CALFRAST (Calibrated finger rub auditory screening test)[5]:** In a quiet room, examiner and patient seats comfortably on chairs. Patient is told to close his eyes. At about 6–10 inches distance nose to nose, examiner fully extends his arms and make finger rubs, with a distance of about 70 cm from the patient. First a stronger rub (CALFRAST: Strong 70), and then a faint rub (CALFRAST: Faint 70) is made. If the patient hears the faint rub, then the test is complete. If the patient does not hear even the stronger rub, then examiner flexes his elbows to 90 degrees, making the distance to about half, i.e. 35 cm. If the patient does not hear at this level, then distance is further reduced to 10 cm, and then to 2 cm. Sensitivity of this test is about 90%.

5. **Tuning fork tests: Air conduction (AC):** In a quiet room, strike a tuning fork of 512 Hz, place near the ear, and compare it on the opposite side. Tuning fork should be held in gentle motion to avoid null points. After the patient points ceasing of the sound, examiner can bring it near his own ear and listen to verify it.

   Bone conduction (BC): In a quiet room, first teach the patient what is bone conduction sound. He should not mistake it with vibration. Strike the tuning fork (512 Hz) and place it on the mastoid tip. When the patient confirms the voice becoming ceased, place it on your mastoid tip and listen. Do this on both sides. This test is also called Schwabach test.

6. **Rinne test:** This test is used to differentiate between conductive and sensorineural hearing loss. It can be done in two ways. In a quiet room, strike the 512 Hz tuning fork and place it at the patient's mastoid tip. As soon as patient stops hearing it, place it against the ear canal. Vice versa can also be done. Normally the AC is twice as long as BC. Rinne is considered positive when AC>BC, and is seen in either normally or CHL. Rinne is considered negative when AC<BC, and is seen in SNHL.

7. **Weber test:** In a quiet room, explain the procedure to the patient. 512 Hz tuning fork is placed either on the vertex of the patient, or nasion, or forehead. The sound is heard through bone conduction. Normally it is heard equally in both the ears. In CHL, the sound is better heard in the worst ear, due to loss of masking effect of air conduction. In SNHL, sound is better heard in the normal ear.

Reflexes: These are useful for testing in infants or children.
Auditory-palpebral reflex: On exposure of the loud sound, there is reflex blinking.

Cochleo-pupillary reflex: Exposure of loud sound produces dilation, or contraction followed by dilation of the pupils.

Auditory oculogyric reflex: Exposure of sound produces eye deviation towards it.

8. **Vestibular nerve examination:** *See* chapter on vertigo.

9. **Other cranial nerves:** Fundus should be examined to rule out raised ICP. Examination of 5th, 9th, 10th, 11th, and 12th cranial nerves should be done to rule out brainstem and CP angle lesions.

Investigations

1. **Audiometry:** Pure tone audiometry and speech audiometry. Pure tone audiometry can be used to test both air and bone conduction. Central causes of hearing loss will have normal pure tone audiometry. A CP angle lesion characteristically produces high frequency sensorineural hearing loss.

2. **Brainstem auditory evoked response (BAER)/auditory evoked potential (AEP):** The potential produced by auditory stimuli is recorded with EEG electrodes. Normally, seven

waves are produced by the auditory structures, which are as follows: (pneumonic – ECOLI MA)

a. **Wave I**—produced by eighth nerve

b. **Wave II**—produced by cochlear nuclei

c. **Wave III**—produced by superior olive

d. **Wave IV**—produced by lateral geniculate body

e. **Wave V**—produced by inferior colliculus

f. **Wave VI**—produced by medial geniculate body

g. **Wave VII**—produced by auditory radiations

A delay between waves I and III, or I and V is seen in CP angle lesions. As the tumour grows in size, all waves after wave I may become unobtainable. Other uses of BAERs are in demyelinating diseases, coma, and brain death.

3. **Tone decay:** At 4000 Hz frequency, 5 dB intensity above the patient's hearing threshold is presented for 60 seconds. Normally a patient hears it. Tone decay up to 5 dB may be seen in CHL. Mild tone decay (5–10 dB), and moderated tone decay (10–25 dB) is seen in cochlear pathology, while severe tone decay (>25 dB) is seen in retrocochlear pathology.

4. **CT and MRI head:** They are ordered when an intracranial lesion is suspected. For a vestibular schwannoma both are required. CT head to see for widening of internal auditory canal (IAC), and distance between IAC and jugular foramen; and MRI brain for characterising the lesion.

Differential Diagnosis

1. **Conductive hearing loss:** It is due to wax, otitis externa or media, foreign body, tympanic membrane perforation, Eustachian tube obstruction, or ossicular chain abnormality (otosclerosis).

2. **Sensorineural hearing loss:** It is due to either cochlear disease like Meniere's disease, trauma, otitis interna, presbycusis, congenital rubella; or eighth cranial nerve lesions like vestibular schwannoma, trauma, infections, toxins, drugs, presbycusis; or due to nuclear lesions like infarcts, MS, inflammations (tuberculosis), neoplasms.

3. **Central hearing loss:** Due to affection of central pathways. It is rare because bilateral lesions have to occur to show any deficit.

REFERENCES

1. Noise-Induced Hearing Loss. https://www.nidcd.nih.gov/health/noise-induced-hearing-loss. Published 2014.

2. Campbell WW. *DeJong's THE Neurologic Examination*. Seventh. Philadelphia: Lippincott Williams & Wilkins; 2013.

3. Sharp JF, Wilson JA, Ross L, Barr-Hamilton RM. Ear wax removal: a survey of current practice. *BMJ*. 1990;301(6763):1251–1253.

4. Roeser RJ, Ballachanda BB. Physiology, pathophysiology, and anthropology/epidemiology of human earcanal secretions. *JOURNAL-AMERICAN Acad Audiol*. 1997;8:391–400.

5. Torres-Russotto D, Landau WM, Harding GW, Bohne BA, Sun K, Sinatra PM. Calibrated finger rub auditory screening test (CALFRAST). *Neurology*. 2009;72(18):1595–1600.

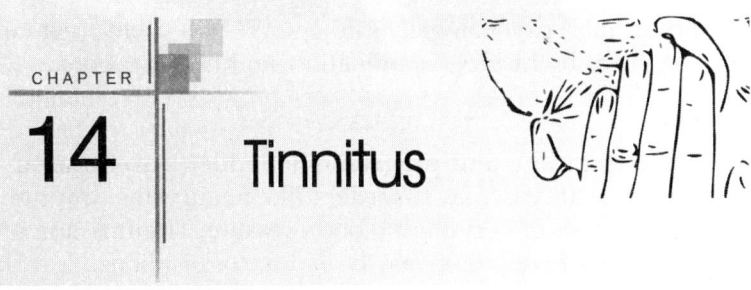

14 | Tinnitus

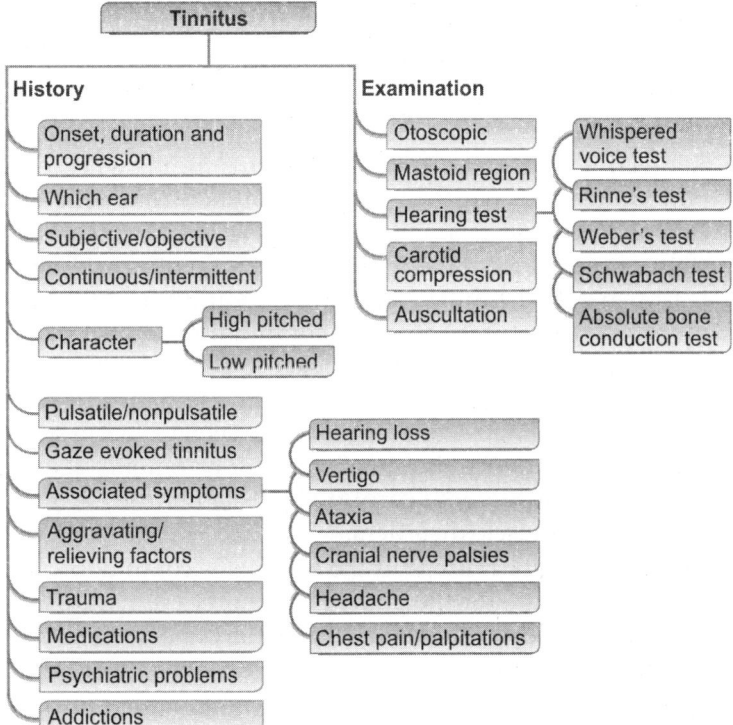

Definition: Tinnitus is the conscious perception of sound when no external stimulus is present. It is derived from the Latin word *tinnire* meaning to ring.

Introduction: Tinnitus is a common complaint with almost everyone having an episode of it, at some point in his or her life. Broadly, it may be because of otologic, neurologic, cardiovascular, psychologic, or others. We will try to reach to the diagnosis by history, examination, and investigations.

History

- **Onset, duration, and progression:** Sudden onset tinnitus may occur in vascular disorders like aneurysms, trauma, sudden onset of very high blood pressure. Tinnitus due to vestibular schwannoma may be sudden or insidious. Rest of the causes are generally insidious ones.

  Progression depends upon the pathology. The CP angle mass (vestibular schwannoma) presents with tinnitus only up to the time until hearing is intact; as the hearing is lost, the tinnitus also disappears.

- **Which ear:** Ask whether it is from one ear (right/left) or both.

- **Subjective/objective:** Subjective tinnitus means when only patient hears it, while objective tinnitus means when an examiner can also hear either through naked ear or through stethoscope. Most of the time subjective tinnitus is present. It is also called tinnitus aurium, non-auditory and non-vibratory tinnitus. It is most commonly because of ear problem—it is argued that hearing loss cause hyper-sensitivity of the neurons (a homeostatic mechanism) which causes tinnitus. Objective tinnitus has been called pseudo-tinnitus or vibratory tinnitus. It is most commonly seen in vascular disorders like carotid stenosis, middle ear bone condition or muscle disorders.

- **Continuous/intermittent:** If the tinnitus is continuous, it is very problematic for the patient. Sometimes it disturbs the patient from sleep. Otherwise, it generally appears when no environmental noise is there.

- **Character:** Ask character of the sound. If the sound heard is high pitch (whinning, hissing, ringing, whistling) it is of sensorineural cause, while if it is of low pitch (roaring, whooshing, wind/waves), it is of conductive cause.

- **Pulsatile/Nonpulsatile:** Pulsatile tinnitus is defined as hearing of the tinnitus coinciding with own pulse. It is because of vascular pathology, and may occur in carotid stenosis, glomus tumours, arteriovenous malformations, or sometimes with raised ICP (pseudotumour cereberi).[1]

Nonpulsatile tinnitus is due to myogenic pathologies, and may occur due to palatal or middle ear myoclonus (palatal microtremor).[2]

- **Gaze evoked tinnitus:** Patient sometimes complain of tinnitus in a particular gaze. It is caused by abnormal connections between vestibular nuclei and cochlear nuclei.
- **Associated symptoms:** Other symptoms like **hearing loss** should be asked. As mentioned it is the most common cause of tinnitus. Patient should be asked about conductive hearing loss like external ear infection, acoustic shock, loud voice listening (especially through earphones), ear wax or effusion. Patient should also be asked about sensorineural hearing loss conditions like excessive loud noise, symptoms of presbycusis (decreased hearing in old age), Meniere's disease (hearing loss, tinnitus and vertigo), vestibular schwannoma (hearing loss, tinnitus and vertigo). Tinnitus is said to be the neuroplastic response of hearing loss.
  - **Ataxia** may be present when there is involvement of vestibular apparatus, cerebellum or brainstem compression.
  - History of **cranial nerve palsies** should be asked.
  - **Headache** may suggest intracranial pathology.
  - **Chest pain/palpitations** may point towards any cardiac pathology. **Hypertension** is also an important cause of tinnitus.
  - **Hyperacusis** is a common associated symptom with tinnitus, noted in 40% of patients.[3]
- **Aggravating and relieving factors:** Vascular tinnitus may be temporary attenuated by carotid compression. Sometimes, patient may experience change in the sound with change in posture of head.
- **Trauma:** Tinnitus may be because of both direct trauma to head and neck, as well as adverse effects of drugs used in head injuries.
- **Medications:** Ototoxic drugs may cause tinnitus. Implicated ones are antibiotoics like polymixin B, erythromycin, vancomycin; antineoplastic drugs like vincristine, mechlorethamine; diuretics like furosemide, bumetanide; aspirin at high doses; quinidine; discontinuation of benzodiazepines.

- **Psychiatric problems:** Anxiety, depression, fatigue are a few comorbidities associated with tinnitus. Serotonin reuptake inhibitors have been used in depressed patients having tinnitus, however, these medications have not been found useful in non-depressed patients.[4]
- **Addictions:** Excessive alcohol, smoking may cause tinnitus.

Examination

- **Otoscopy:** Exclude presence of wax, tympanic membrane perforation, foreign bodies, blood and exudates.
- **Examination of mastoid region** for swelling and tenderness.
- **Hearing examination:** These can identify whether there is hearing loss present, and if present, whether it is of conductive or sensorineural type. Whispered voice test/ticking watch test/finger rub, Rinne's test, Weber's test, Schwabach test and absolute bone conduction test to differentiate CHL from SNHL, should be performed. (Explained in detail in Chapter of Hearing Loss.)
- **Carotid compression test:** It may inhibit the vascular tinnitus caused by carotid stenosis.
- **Auscultation:** Pulsatile tinnitus can be detected with it.

Investigations

No objective tests are available for tinnitus.

1. **Audiometry:** It is an objective test to measure the hearing loss at different frequencies, and characterize it to be conductive or sensorineural. High frequency sensorineural hearing loss is seen in retrocochlear lesions like vestibular schwannoma.
2. **CT head:** It is the investigation of choice in cases of trauma. Temporal bone fractures, space occupying lesions, petrous bone erosions, cochlear abnormalities, etc. are well seen in it.
3. **MRI brain:** It is valuable for intracranial pathologies. Contrast can be added if SOLs or inflammations are suspected.
4. **USG Doppler neck:** It is done when pulsatile tinnitus is the complaint.
5. **Angiography** (conventional, CT angiography, or MR angiography): It is also indicated when pulsatile tinnitus is the complaint. Pathologies like carotid stenosis, AVMs, glomus tumours, etc. can be identified.

Table 14.1: Differential diagnosis of tinnitus[5]			
S. No.	System	Pathologies	Types
1.	Ear	Infections	Otitis media, labyrinthitis, mastoiditis
		Neoplastic	Cholesteatoma
		Labyrinthine	Meniere's disease, semicircular canal pathologies
		Others	Impacted wax, otosclerosis, presbycusis, noise exposure (most common), ototoxic drugs
2.	Neurological	Inflammations	Meningitis
		Neoplastic	Vestibular schwannoma, CP angle meningiomas, glomus tumours
		Vascular	Carotid stenosis, AVMs
		Others	Epilepsy, migraine, multiple sclerosis
		Trauma	Temporal bone fractures
3.	Orofacial	Joint abnormalities	TMJ disorders
		Vascular	Giant cell arteritis
4.	Cardiovascular	Hypertension	
5.	Immunologic	SLE, systemic sclerosis	
6.	Rheumatologic	Rh arthritis	
7.	Endocrine	Hormonal imbalances	DM, hypothyroidism, pregnancy changes
8.	Psychological		Anxiety, depression, post-traumatic stress disorder
9.	Idiopathic		
10.	Drugs		Antibiotoics like polymixin B, erythromycin, vancomycin; antineoplastic drugs like vincristine, mechlorethamine; diuretics like furosemide, bumetanide; aspirin at high doses; quinidine; discontinuation of benzodiazepines.

REFERENCES

1. Stouffer J, Tyler RS. Characterization of tinnitus by tinnitus patients. *J. Speech Hear. Disord.* 1990;55(3):439–453.
2. Bhimrao SK, Masterson L, Baguley D. Systematic review of management strategies for middle ear myoclonus. *Otolaryngology—Head and Neck Surgery.* 2012;146(5):698–706.
3. Anari M, Axelsson A, Eliasson A, Magnusson L. Hypersensitivity to sound: questionnaire data, audiometry and classification. *Scand. Audiol.* 1999;28(4):219–230.
4. Robinson SK, Viirre ES, Bailey KA, Gerke MA, Harris JP, Stein MB. Randomized placebo-controlled trial of a selective serotonin reuptake inhibitor in the treatment of nondepressed tinnitus subjects. *Psychosom. Med.* 2005;67(6):981–988.
5. Baguley D, McFerran D, Hall D. Tinnitus. *The Lancet.*382(9904):1600–1607.

15

Vertigo

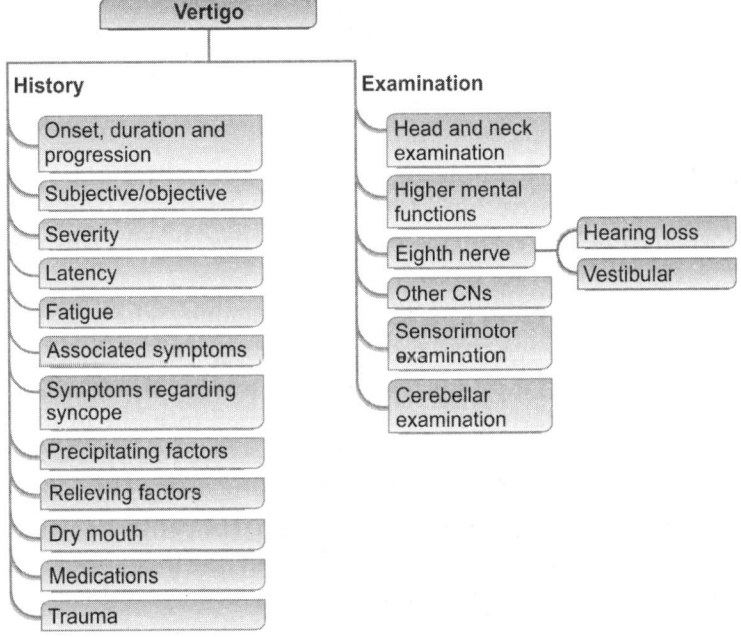

Definition: Vertigo is when a person feels like he is moving when actually he is not.[1]

Classification: It may be classified as **peripheral or central vertigo**. The pathology in the peripheral vertigo lies in the inner ear or the vestibular system, which constitutes semicircular canal, vestibule and the vestibular nerve. The pathology in central vertigo lies in the brainstem or cerebellum.

History

- **Onset, duration and progression:** Sudden onset of vertigo occurs in trauma, CVA in cortex, brainstem, cerebellum and labyrinth. Other causes of sudden onset and short duration include labyrinthitis. Insidious onset ones with long duration include BPPV, Meniere's disease, CP angle masses, migraine and age related vertigo.

- **Subjective/objective:** In subjective vertigo patient feels that they are moving, while in objective vertigo the patient feels that the surroundings are moving.[2] There is a third type called pseudovertigo in which person feels that there is intense rotation inside his head.

- **Severity:** The severity is less in the central vertigo.

- **Latency:** Peripheral vertigo is of short latency, while central vertigo is of long latency.

- **Duration of each episode:** The episodes of BPPV last less than one minute.[3] In Ménière's disease the episodes of vertigo lasts more than twenty minutes. In labyrinthitis vertigo can last for days.

- **Fatigue:** Peripheral vertigo is fatigueable, while central vertigo is not.

- **Associated symptoms: Nausea, vomiting, tinnitus and decreased hearing** are common accompaniment in peripheral causes of vertigo. **Facial weakness** localises the lesion to be in the cerebellopontine angle or brainstem.

  Slurred speech, diplopia are features of central vertigo. **Dysphagia** may be associated with both due to involvement of lower cranial nerves, although more with central vertigo. **Ataxia** may occur both because of vestibular system and cerebellar involvement. Ataxia increasing with eye closure or in darkness point towards the vestibular origin. **Headache** may occur due to raised ICP, or due to dural irritation by the posterior fossa masses.

- **Symptoms regarding syncope:** Disequilibrium should be differentiated with syncope in which blackouts, palpitations, chest pain, shortness of breath, pallor and sweating are predominant features. Anaemia, convalescence, arteriosclerosis may be associated when syncope accompanies vertigo.

- **Precipitating factors:** Vertigo on prolonged spinning of head is physiologic to everybody. Change in head positions may precipitate vertigo in BPPV, and patients are normal in the intervening period.[4] Movements of neck may precipitate vertigo in cervical vertigo also. Similarly, diplopia may precipitate vertigo in ocular vertigo.
- **Relieving factors:** Repositioning (Epley's manoeuvre) relieves vertigo in BPPV; antihistaminic drugs helps in peripheral vertigo.
- **Dry mouth** can be present with dizziness due to dehydration.
- **Medications:** Alcohol, aspirin, vestibulotoxic drugs like aminoglycoside antibiotics may produce vertigo.
- **Trauma** may produce vertigo which is transient and passes off as the head is maintained in the static position.

Examination

- **Head and neck examination:** Vertigo patients have frequently other ear, nose and throat pathology. Common conditions, which should be sought for, are cerumen impaction in ear, otitis media, sinusitis, oropharyngeal findings, and visual acuity.
- **Higher mental functions:** If suspicion is of central vertigo and cortical dysfunction is suspected, these should be tested for.
- **Eighth nerve examination:**
    - **Hearing examination:** *See* chapter on hearing loss
    - **Vestibular examination:** These include tests to detect nystagmus, vestibulo-ocular reflexes, and vestibulospinal reflexes.

  Nystagmus: Horizontal nystagmus is associated with peripheral vertigo. Sometimes patients of BPPV report of nystagmus with the change in head positions, while labyrinthitis patients may have nystagmus without relation to head position.[4] Vertical or torsional nystagmus is associated with central vertigo.
- **Spontaneous nystagmus:** Place the patient in primary gaze and see for spontaneous nystagmus, repeat with Fresnel lenses (Fig 15.1). If nystagmus occurs, see for fast and slow phases.

Fig. 15.1: Fresnel goggles containing Fresnel lenses: These are convex lenses of +20 magnification power. An illumination system is also present. Room lights are darkened. Patient's nystagmus can be well seen as they are illuminated and there is absence of fixation due to darkness in the room.

- **Gaze nystagmus:** Ask the patient to look at a gaze of about 20–30 degrees from the primary position and see for appearance or change in the nystagmus.
- **Smooth pursuit:** Make the patient to see and follow the finger up-down, right and left (not more than 60 degrees), see for the nystagmus.
- **Saccades:** Ask the patient to see two fingers back and forth held twelve inches apart in vertical and horizontal plane. Observe for latency, speed, accuracy and conjugate movements. Delayed saccadic movements are because of cortical (frontal lobe) and brainstem disease (PPRF).
  Tests for vestibulo-ocular reflex (VOR): These test for integrity of vestibular and occulomotor systems as well as their connections.
- **Dynamic visual acuity:** Ask the patient to read Snellen's chart while moving the head. Normally, with intact VOR he/she can read. Abnormal test is a sign of vestibular dysfunction.
- **Caloric reflex test:** This method was developed by Robert Barany, who got Nobel prize in 1914. Technique: Ask the patient to lie down on bed. Elevate the head by 30 degrees (this makes the horizontal semicircular canal parallel to the ground). Instil warm water (44°C) in one ear, and watch for nystagmus. Due to warm water the endolymph will rise

creating the signals as if head is moving towards its direction. This will lead to slow movements of eyes contralaterally, and fast nystagmus ipsilaterally. Reverse happens with cold water (30°C). Mnemonic **COWS:** Cold opposite, warm same. Absence of nystagmus means weakness or absence of activity of vestibular system of that side.

- **Doll's eye or oculocephalic reflex:** A cervical problem is ruled out first. Ask the patient to lie down. Stand behind the patient holding his head with both hands. Turn his head alternatively to right and left, watching his eyes. Normally, with intact VOR eyes move opposite to the head in order to maintain the gaze. Absence of it indicates vestibular dysfunction. It is especially helpful in comatose patients.

- **Head thrust test:** Patient is asked to fix gaze at an object. Examiner moves his head sideways. Absence of gaze fixation, with lagging of eyes behind, and rapid compensatory movement is seen in vestibular dysfunction.

Testing for Vestibulospinal Reflex (VSR)

- **Romberg's test:** Ask the patient to stand with his feet together. First, observe swaying towards any side. Its presence indicates ipsilateral cerebellar disease. If patient is stable, then ask him to close his eyes. Observe for 60 seconds. Stand close to the patient in order to catch him, if he falls. Swaying with foot dislodgement from the ground is considered positive Romberg's test, and indicates impaired proprioception or vestibular system disease. Patient of proprioception disorder may fall towards any side, but a vestibular disorder patient will consistently fall on one side, which will be ipsilateral to the vestibular disease. Absence of the fall is termed negative Romberg's test.[5]

- **Past pointing:** Explain to the patient to touch index finger of the examiner with his index finger with outstretched arm. Now, ask the patient to close his eyes, and repeat the procedure. Progressively, his arm will miss the target and will go away, ipsilateral to the diseased labyrinth.

- **Star walking test:** Ask the patient to walk, with eyes closed, in a straight line forward, turn, and then back again. Progressively he will turn ipsilaterally towards the diseased labyrinth and will make a star.

- **Fukuda's stepping test:** Ask the patient to march, with eyes closed, at a single spot. Progressively, he will turn towards the diseased labyrinth.
- **Gait examination:** Ask the patient to flex and extend the head while walking straight, with eyes open. This will test the superior and posterior semicircular canals. Similarly, ask him to turn his head right or left while walking, which will test the horizontal semicircular canals. Instability during walking will suggest vestibular dysfunction.
- **Other cranial nerves:** Other cranial nerves, especially, 5th and 7th should be examined.
- **Sensorimotor examination:** To look for brainstem disease or compression.
- **Cerebellar examination:** Cerebellar compression can be found in CP angle masses producing vertigo. Similarly, cerebellar pathologies itself can lead to vertigo.

Investigations

- **MRI brain:** These are ordered to see for intracranial pathologies. Space occupying lesions, multiple sclerosis, stroke can be seen on it.
- **MRI cervical spine:** It can be done to rule out cervical spine disorders, if in doubt.
- **CT head:** It is good for visualising fractures, and in patients who cannot undergo MRI.
- **Electro/video-nystagmography (ENG/VNG):** Nystagmus can be better seen using electrodes placed around the eyes, or using video camera installed goggles. This is also performed under dim/dark light to avoid fixation.
- **Audiometry:** It is ordered if hearing loss is found during the examination. It will differentiate between conductive and sensorineural hearing loss, and will tell for what frequencies.
- **Auditory brain stem responses (ABR):** Its use has declined, but can be of use in patients in which MRI cannot be performed (e.g. metal implants), or audiometry cannot be performed (like in infants).

Differential Diagnosis

- **Physiological:** Sea sickness, motion sickness, etc.
- **Pathological:** Causes of peripheral vertigo

- Ear diseases: External ear like wax, foreign body; middle ear diseases like otitis media, otosclerosis, Eustachian tube blockage; inner ear diseases like infections, drugs (salicylates, quinine, streptomycin), haemorrhage, Meniere's disease.
- Vestibular nerve disorders like trauma, vestibular schwannoma, meningioma compressing it, neuronitis
- BPPV
- Labrynthitis
- Superior canal dehiscence syndrome
- Age related (because of age related nerve conduction slowing)
- Motion sickness

• **Causes of Central vertigo:**
- Cerebellar neoplasms
- Cervical spine disorders
- Stroke
- Multiple sclerosis
- Head trauma affecting cortex
- Migraine
- Epilepsy
- Chiari malformation
- Multiple sclerosis
- Parkinsonism

REFERENCES

1. Post RE, Dickerson LM. Dizziness: a diagnostic approach. *Am Fam Physician*. 2010;82(4):361–368.
2. Berkow R, Fletcher A. Merck Manual of Diagnosis and Therapy, Merck and Co. Inc., Rahway, NJ. 1992.
3. Hogue JD. Office evaluation of dizziness. *Primary Care: Clinics in Office Practice*. 2015;42(2):249–258.
4. Kerber KA. Vertigo and dizziness in the emergency department. *Emergency medicine clinics of North America*. 2009;27(1):39–50.
5. Lanska DJ, Goetz CG. Romberg's sign development, adoption, and adaptation in the 19th century. *Neurology*. 2000;55(8):1201–1206.

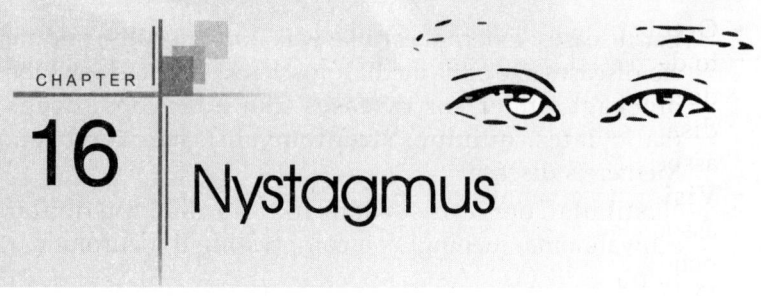

16 Nystagmus

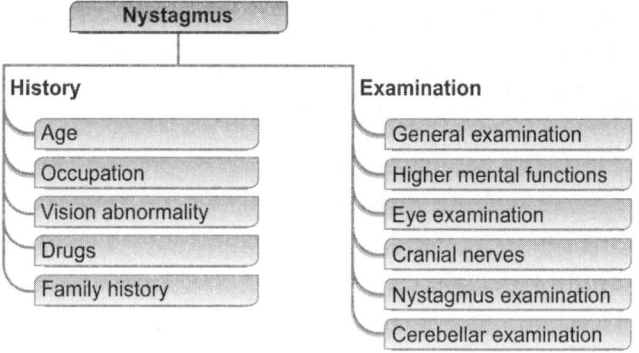

INTRODUCTION

Nystagmus is defined as involuntary oscillatory movements of the eyeballs. Steady gaze requires—fixation, vestibulo-ocular reflex, and a gaze holding system (neural integrator) during eccentric gaze.[1] The direction of the nystagmus is labelled by its fast component.

History

- **Age**
  - *Congenital nystagmus:* Since birth. It is mostly horizontal, and remain horizontal even in upgaze and downgaze. Paradoxical reversal of nystagmus direction is pathognomic feature.
  - *Spasmus nutans:* 6 months to 12 months of age. Classic triad is nystagmus, head nodding and torticollis. Nystagmus is low amplitude, high frequency and dysconjugate.
  - Other causes of infantile nystagmus include aniridia, albinism, congenital cataract, etc.

- **Occupation:** Miners may present with nystagmus. It is due to decreased illumination and characterised by nystagmus, decreased vision, photophobia, and night blindness.[1] It disappears after cessation of mining. Plumbism is also associated with nystagmus.
- **Vision abnormality:** Decreased visual acuity may be associated with nystagmus. Amblyopia may present with pendular and fixation nystagmus in the early life.
- **Drug history:** Drug intoxication is the most common cause of nystagmus. Included in it are alcohol, phenytoin, barbiturates, aminoglycosides.
- **Family history:** It may be familial, and is then present since birth.

Examination

- **General examination:** Examine for blood pressure, and signs of atherosclerosis.
- **Higher mental functions:** These should be examined to rule out central from peripheral causes of nystagmus.
- **Examination of eye:** Examine for opacities, refractive errors, and ophthalmoplegia.
  Eyes should be examined first in the primary gaze, and then in upward, downward, and lateral gazes.
- **Cranial nerves:** Examine second, third, fourth and sixth nerves which may cause ptosis, diplopia, or squint and lead to nystagmus. VIIIth nerve examination should be done to rule out vestibular abnormalities.
- **Nystagmus examination:**
  - **Frequency:** Establish whether the nystagmus is fine or coarse. In irritative lesions like vestibular schwannoma: Coarse nystagmus occurs towards and fine opposite the lesion **(Bruns nystagmus)**. Congenital nystagmus is fine.
  - **Latency:** Notice after how much time the nystagmus starts after head shaking. Latency of up to 30 seconds may occur in peripheral lesion while it is absent in central lesions.
  - **Fatigue:** Peripheral lesions are fatigueable while central lesions are not fatigueable.
  - **Characterize the type of nystagmus:** Note the fast component, slow component, whether gaze evoked/paretic, direction, etc. to characterize and classify the nystagmus.

- **Cerebellar examination:** Perform detailed cerebellar examination to rule out its pathology or involvement.

Investigations

1. **MRI brain:** This is ordered whenever an intracranial cause is suspected.
2. **Ophthalmoscopy and slit lamp examination:** These are needed when corneal, vitreal or retinal cause is suspected (Tabel 16.1).

Table 16.1: Clinical types and differential diagnosis[2-4]		
Nystagmus type	*Characteristics*	*Localisation*
Upbeat nystagmus	Eyes beat superiorly in primary gaze	Cerebellar vermis, medulla
Downbeat nystagmus	Eyes beat inferiorly in primary gaze	Cervicomedullar junction.
Convergence retraction nystagmus	Convergence and/or retraction of eye globes	Rostral midbrain, pretectum, pineal, posterior IIIrd ventricle
Rebound nystagmus	Eyes beat horizontally in opposite direction on coming to primary gaze	Cerebellum or its connections
Periodic alternating nystagmus	Cyclical nystagmus. Eyes beat horizontally in one direction for 1–3 min, pauses, and then in opposite direction	Brain stem or cerebellum
See saw nystagmus	One eye rises and intorts, and other eye falls and extorts. This then repeats reversely.	Suprasellar, chiasmal, and anterior third ventricle
Ocular bobbing	Eyes jerk in downward direction and then slowly drifts back to normal	Pons
Ocular flutter	Rapid repeated horizontal saccades	Cerebellum (dentate nucleus) or its connections
Opsoclonus	Continuous chaotic saccades	Cerebellum (dentate nucleus) or its connections

Contd.

Table 16.1: Clinical types and differential diagnosis[2-4] (Contd.)

Nystagmus type	Characteristics	Localisation
Gaze evoked	Nystagmus evoked by lateral gaze fixation (Not far lateral gaze)	Brain stem lesion
Gaze paretic	Nystagmus evoked by lateral gaze fixation towards paretic muscle	Incomplete CN palsies
Vestibular nystagmus	Physiological nystagmus evoked by head rotation, instilling ward or cold water in the ears	Absent in vestibular nerve or vestibular nuclei lesions.
Optokinetic	Physiological nystagmus seen when eyes follow an object in a set pattern and suddenly fixate to the following set.	Absent unilaterally towards ipsilateral cerebral lesion

Classification and Differential Diagnosis[2-4]

1. True Nystagmus

a. **Pendular:** There is rhythmic side to side movement of the eyeballs. It occurs only when the eyes are in primary gaze at rest. Causes include **ocular** ones like corneal/vitreal opacities, congenital cataract, ophthalmia neonatarum; **retinal** ones like chorioretinitis, miner's nystagmus, colour blindness; and **central** ones like albinism, spasmus nutans.

b. **Jerky:** The nystagmus is non-rhythmic and faster in one direction. Causes include:

   i. **Peripheral:** Features are similar to peripheral vertigo, like fatigueability, short latency, more intensity, and possibility of associated tinnitus and hearing loss. Causes include:

   Ocular ones like ophthalmoplegias, polyneuritis, myasthenia gravis, diphtheria, manganese poisioning. Associated muscle paralysis will be seen in ophthalmoplegias, diurnal variation will be seen in myasthenia gravis. **Gaze paretic nystagmus** is seen in incomplete nerve palsies. Unilateral lateral rectus palsy may lead to **dissociate nystagmus** (nystagmus in only one eye).

Vestibular ones include labyrinthitis, occlusion of internal auditory artery, tumours like vestibular schwannoma.

Cervical cord lesion:[5] Nystagmus in cord lesions may occur due to involvement of vestibular tracts. In syringomyelia and syringobulbia a **rotatory** variety of nystagmus is seen.

ii. **Central:** Features are lack of fatigueability, mild intensity, long latency, and absence of tinnitus, vertigo or hearing loss. Aetiology includes:

Brain stem lesions like tumours, disseminated sclerosis, vascular lesion, tuberculomas. There will be associated other cranial nerve palsies, long tract signs. There will be coarse unidirectional nystagmus of either horizontal or vertical. **Vertical upbeat** nystagmus is seen in tegmentum lesions and medulla, and **downbeat nystagmus** is seen in Chiari malformations, lesions of vestibule, cerebellum, and the medulla. **Convergence retraction nystagmus** is characterised by convergence and retraction of the eye globe, and is seen in rostral midbrain, pretectum, posterior commissure, and posterior third ventricle. **Ocular bobbing** is characterised by downward jerking movements followed by slow drift back to primary position is linked to pontine haemorrhage or infarcts. Internuclear ophthalmoplegia leads to **dissociate nystagmus.**

Cerebellum lesions like tumours, PICA infarcts, hereditary ataxias. Nystagmus is more marked on looking towards side of lesion. Associated cerebellar features like dysmetria, dyssynergia, etc. may be present. **Rebound nystagmus** characterised by horizontal movements beating to opposite side on returning to primary position, and **periodic alternating nystagmus** characterised by cyclical beating to one side and then other side for 1–3 minutes is linked to cerebellum disease or its connections. **Opsoclonus** characterised by continuous chaotic saccades in any direction is seen with lesions of dentate nucleus, cerebellar connections and brainstem. Normal saccadic ability is also impaired in cerebellar lesions.

Cortical lesions like tumours, vascular lesions, inflammations, epilepsy; or hysteria. Lobar features of specific lobes may be found. For example, epileptic nystagmus has been reported from parieto-occipital region,[6] and frontal and temporal regions.[7]

Anterior third ventricle or parasellar region like tumours (craniopharyngioma, etc.), trauma, septo-optic dysplasia, etc. may have **see-saw nystagmus** characterised by pendular nystagmus where one eye intorts and other extorts, and then reverse occurs.

c. **Infantile:** These are found in the infantile age group, due to causes like albinism, optic nerve hypoplasia, and congenital cataracts. It may be either congenital or manifest latent nystagmus.

2. Pseudonystagmus/ Physiological Nystagmus

a. **Optokinetic nystagmus:** Normally a person experiences nystagmus when he looks to a series of rapidly passing by objects: Like a rotating drum, objects passing by when viewing through a train/car. In this nystagmus slow phase occurs towards the moving objects (from parietal lobe), while fast saccadic movement occurs opposite the direction of the moving objects (by frontal lobe). Parietal lobe lesion patients thus will not be having this nystagmus, while frontal lobe lesion patients will have opposite the normal response.

b. **Vestibular nystagmus:** It occurs during rotation of head, even in darkness. It is due to labyrinthine connections to the vestibular nuclei and cerebellum. It can be induced by instilling warm or cold water into the ear canal. Nystagmus is horizontal, torsional or oblique depending upon the position of the head. When the head is raised to 30 degrees, there is nystagmus (fast component) to the opposite side by instilling cold water (30 degrees temperature), and towards same side by instilling warm water (43 degrees temperature).

c. **End point nystagmus:** In normal individuals, there is some amount of nystagmus (<2 degrees) at the extremes of gazes i.e. >40 degrees.

Important Points

- **Alexander's law:** According to this law, in first degree the nystagmus is present in only the eccentric gaze. The second degree nystagmus is present even in primary gaze, and in the third degree, nystagmus is present even when looking in opposite gaze.

REFERENCES

1. http://www.nature.com/nature/journal/v91/n2263/abs/091030a0.html
2. Campbell WW. *DeJong's THE Neurologic Examination*. Seventh. Philadelphia: Lippincott Williams & Wilkins; 2013.
3. Gami N. *Bedside Approach to Clinical Neurology*. 2nd ed. Jaypee; 1998.
4. Abadi RV. Mechanisms underlying nystagmus. *JR Soc Med*. 2002;95(5):231–234.
5. Booth CB. Does nystagmus occur in lesions of the cervical cord? AMA Arch Neurol Psychiatry 1952;67: 69–71.
6. Kellinghaus C, Skidmore C, Loddenkemper T. Lateralizing value of epileptic nystagmus. *Epilepsy Behav*. 2008;13(4):700–702. doi:10.1016/j.yebeh.2008.07.015.
7. Kim K-S, Kim YH, Hwang Y, Kang B, Kim DH, Kwon YS. Epileptic Nystagmus and Vertigo Associated with Bilateral Temporal and Frontal Lobe Epilepsy. *Clin Exp Otorhinolaryngol*. 2013;6(4):259–262. doi:10.3342/ceo.2013.6.4.259.

Dysphagia

```
┌─────────────────────┐
│      Neck pain      │
└─────────────────────┘
```

History

- Onset, duration and progression
- Age
- Severity
- Charcter
- Location and radiation
- Aggravating and relieving factors
- Diurnal variation
- Restriction of neck movement
- Sensori-motor problems in limbs
- Shoulder movement
- Chest/epigastric pain
- Inter-scapular pain
- Other
- Trauma
- Occupation
- Treatment and present status

Examination

- General
- Respiratory
- Face
- Oral, pharynx and larynx
- Neurologic examination
- Other clinical tests

- Shortness of breath
- Decreased sensation
- Swallowing difficulty/ nasal regulation
- Bowel and bladder involvement
- Drooping of eyelids, decreased sweating
- Constitutional symptoms

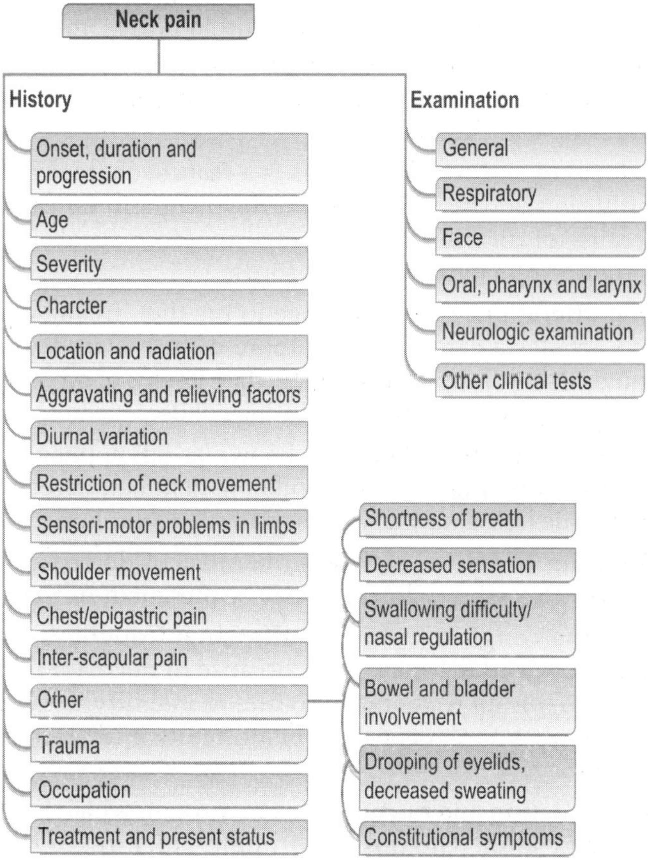

Definition: Difficulty in swallowing.

There are generally two causes of dysphagia: Neurogenic and gastro-intestinal (GI). Symptoms and signs are mostly similar. For neurologists and neurosurgeons, the task is first to make the diagnosis of CNS disease, and then to look for dysphagia to prevent the associated morbidities. In the neurologic dysphagia, we have to characterize it in between degenerative and non-degenerative causes.

History

- **Onset, duration and progression:** Acute onset dysphagia is common because of inflammation or esophageal strictures due to corrosive poisoning. In the category of neurogenic dysphagia stroke, trauma has acute onset, while degenerative causes like PD have gradual onset. Progressive dysphagia can be because of strictures, achalasia cardia, medullary paralysis, degenerative neurological disorders, and neoplastic lesions.

- **Relapsing and remitting:** This is a feature characteristic of multiple sclerosis. Intermittent dysphagia in GI causes is because of diffuse esophageal spasm, esophageal rings, or chronic esophageal motility disorder.

- **Diet alteration:** Ask for change in the diet. In vagal palsy and achalasia cardia there is more difficulty in degluting liquids than solids; while in obstructive lesions of oesophagus, there is more difficulty in degluting solids.

- **Pain:** Pain during deglutition, known as odynophagia, may point towards inflammatory conditions of throat. Oesophageal neoplastic lesion may also cause painful deglutition.

- **Drooling of saliva:** May occur because of decreased deglutition. Excessive sialorrhoea may be present in ALS and PD.

- **Choking/coughing while swallowing:** Coughing during eating or drinking may occur because of decreased deglutition. Aspiration pneumonia, due to inability to swallow secretions, is a frequent cause of morbidity in a neurological patient.

- **Regurgitation and burping:** Ask in through which nostril— single or both. The nostril through which regurgitation occurs has ipsilateral vagus nerve weakness. Excessive burping is a feature of Huntington's disease.

- **Speech problem:** Nasal twang or hoarseness of voice may point towards vagal (recurrent laryngeal nerve) palsy.

- **Recurrent chest infection:** It points dysphagia. Aspiration pneumonia due to swallowing dysfunction is a common cause of morbidity and mortality of many neurologic illnesses like stroke, head injuries, HD.[1]
- **Weight loss:** Recent weight loss may occur due to chronically decreased dietary intake.
- **Oral problems:** Ulcerations due to inflammation, food residue in dementia, and tongue tremors in PD may be seen.
- **Postural variation:** History of regurgitation of undigested food while lying down, with cough at night may point towards hypopharyngeal diverticulum.
- **Headache:** It may point toward an intracranial pathology.
- **Cognition:** Patients of dementia have frequent association with feeding abnormalities leading to dysphagia. Many such patients are unaware of their disease leading to silent aspirations.
- **Exclude** confounding symptoms

  Masticatory difficulty: Its because of fifth nerve palsy.

  Tongue movements: Its difficulty indicates hypoglossal nerve weakness.

  Ulcerative lesions of mouth
- **Drug history:** Tardive dyskinesias due to neuroleptic drugs, like haloperidol, may present as dysphagia.
- **Past history:** Enquire about history of stroke,[2] trauma,[3] PD,[4] motor neuron disease and myopathy. On a therapeutic note, a stroke patient should be assessed for dysphagia by a speech therapist, as delay in assessment may lead to increase chances of stroke associated pneumonia.[5]
- **Treatment taken and present status:** Patients of oral and tongue cancers, and Huntington's disease may benefit from Masako and Mendelson manoeuvre.[6] In the Masako manoeuvre patient protrudes the tongue between the front teeth and swallows in this position. This increases the base of tongue movement contacting with the posterior pharyngeal wall.

  In the Mendelson manoeuvre patient holds the larynx, for about 3 seconds, at the peak of hyolaryngeal elevation. This will prevent the food entering into the larynx.

Examination

- **General examination**: Look for nutrition, built, weight, dehydration. These may be reduced in chronic dysphagia.
- **Respiratory examination:** Coughing and choking may be present. They can be absent in silent aspiration pneumonitis, which also needs to be looked for.

  The normal swallowing consists of four phases—**oral preparatory phase** consisting of chewing and making food bolus, **oral phase** consisting of tongue propelling the food bolus into the pharynx, **pharyngeal phase** consisting of reflexive swallow through the pharynx, and **oesophageal phase** consisting of oesophageal peristalsis propelling the food to stomach. Patients of neurogenic dysphagia suffer from disorders of first three phases. We have to examine in view of each of these phases.

- **Face:** Look for drooling, asymmetry, reduced mastication, atrophy of muscles, nasal regurgitation. Drooling and asymmetry may be present in seventh nerve paresis. Reduced mastication and atrophy of masticatory muscles may be present in fifth nerve paresis. Nasal regurgitation may be present in tenth nerve paresis.

- **Oral, pharynx and larynx:** See for persistent residue in the oral cavity. Rule out ulcerative lesions of mouth. Persistent ulcerative lesions may be because of Hiatus hernia and GERD. In head injuries patient may have decreased oral control and decreased base of tongue retraction. Patients of stroke may have retention of food in the lateral sulci and delayed oral transfer. Patients of dementia may keep the food in the mouth and forget about swallowing. PD patients may have lingual tremors.

  Vocal cords should be examined to see for their position and movements, which may be impaired in recurrent laryngeal nerve palsy.

- **Neurologic examination:** MMSE should be done to look for dementia which may impair feeding. One should look for decreased mentation, aphasia, dysarthria, hemineglect, velar, pharyngeal and laryngeal paralysis, spastic hemiplegia, loss of pain and temperature, and ataxia. Orobuccolingual dyskinesias (repetitive movement of lower face, lips, and tongue) may occur due to neuroleptic drugs.

- **Lower cranial nerve examination:** *It is the most important part of examination.*
  - Examination of uvula and palate at rest, quiet breathing, and phonation: Drooping of palate, and flattening of arch occurs on the side of vagal palsy. On phonation, normally it remains in the midline, but in vagal palsy it deviates towards the normal side. In bilateral vagal palsies: The palate will not be able to elevate on phonation, and median raphae is central.
  - Gag reflex: Palatal gag reflex is absent on the involved side. Presence or absence of gag reflex should be correlated clinically, as its absence may be seen in normal individuals, and may be present in neurogenic dysphagia.
  - Examination of speech: Speech will be having nasal quality even in unilateral vagus palsy.
  - Spontaneous coughing or cough reflex may be impaired
  - Oculocardiac reflex: Bradycardia (minimum 5–8 beats per min) caused by pressure on the eyeball. This may be absent in vagal palsies.
  - Other reflexes which may be absent are: Vomiting, cough, swallowing, hiccup, yawning, carotid sinus reflex.
- **3 Oz test:** This is a clinical test for swallowing. Patient is asked to drink 3 ounces of water in front of the examiner. Any cough, change of voice, or decreased rate of swallowing is noted. This test has a poor sensitivity as many silent aspiration goes unnoticed.
- **Other clinical test:** Daniel and colleagues studied six clinical features—abnormal volitional cough, abnormal gag reflex, dysarthria, dysphonia, cough after trial swallow, and voice change after trial swallow. Any two of these factors, if present, suggest risk of aspiration with 92% accuracy.[7]
- **Bedside tests:** Periprandial pulse oximetry has been used to see the oxygen saturation before, during and after the swallowing. Dropping down of O_2 saturation has been said as to be a predictor of impaired swallowing.

  Tartaric acid inhalation (maximum three inhalations) will cause cough. Absence of cough is said to be a predictor of impaired swallowing.

  Imaging studies: It directly evaluates the physiological status of the swallowing mechanism.

- **Videoflouroscopic swallow study (VSS):** It is said to be the gold standard for dysphagia evaluation. It allows for the testing of oral, pharyngeal and esophageal functioning. Patient is asked to drink a radio-opaque contrast, like barium. Its passage is seen under fluoroscopy.
- **Videoendoscopy:** Its a bedside or office test, in which a patient swallows a drink/food, and examiner visualises it by an endoscope placed in the oral cavity. The spillage of food into the larynx before the swallow can be noted.
- **CT/MRI:** These are ordered to confirm CNS disease.

Differential Diagnosis of Dysphagia

- **GI disorders:** Strictures, rings, diffuse esophageal spasm, achalasia cardia, malignancy
- **Central nervous system disorders[8]:** Degenerative and non-degenerative
  Non-degenerative
  Vascular stroke
  Trauma: Head injury
  Neoplastic: Brain tumours (posterior fossa tumours)
  Congenital: Cerebral palsy
  Iatrogenic: Medication induced (tardive dyskinesia), surgery induced (carotid endarterectomy, cervical spine surgery).

Degenerative

Progressive course

Dementia: Alzheimer's dementia, frontotemporal dementia, Lewy body dementia, vascular dementia.

Movement disorders: Parkinson's disease, progressive supranuclear palsy, olivopontocerebellar atrophy, Huntington's disease, Wilson's disease.

Relapsing-remitting course: Multiple sclerosis

Stroke: It is the most common cause of neurogenic dysphagia. Around 65% of stroke patients suffer from dysphagia with risk of aspiration pneumonitis in around 50%.[9] Neurologic symptoms vary depending upon the location of ischemic infarct. A patient of supratentorial infarct may present with spastic hemiplegia, aphasia, dysarthria, hemineglect, decreased mentation, etc. With a lateral medullary infarct patient may present with ataxia, contralateral loss of pain and temperature, ipsilateral velar, pharyngeal and laryngeal paralysis. The swallowing abnormality may vary and may be at many steps including retention at oral lateral sulci, oral transfer delay, pharyngeal swallowing elicitation delay, hyolaryngeal elevation decrement, and aspiration.

Trauma: TBI features vary depending upon the site and nature of injury. Aspiration pneumonia due to swallowing abnormalities is a common cause of morbidity and mortality associated with it.

Neoplastic lesions: Posterior fossa tumours, whether compressive or infiltrative, may result in dysphagia due to lower cranial nerve palsies or medullary involvement. In addition, iatrogenic surgical or radiation related factors also may result in this manifestation.

Cerebral palsy: Severity of this *in utero* or perinatal insult determines the dysphagia.

Medication induced: Tardive dyskinesia due to neuroleptic drugs, like haloperiol, may lead to dysphagia. Newer drugs like clozapine and risperidone have less of this side effect.

Surgery related: Carotid endarterectomy (CEA), and cervical spine surgery (anterior approach) carries the risk of dysphagia. CEA carries risk of seventh, tenth, and twelfth nerve injury due to their vicinity, and anterior approach to cervical spine carries risk of recurrent laryngeal nerve injury.

Dementias: Alzheimer's dementia, frontotemporal dementia, lewy body dementia, and vascular dementia may lead to dysphagia. Swallowing abnormality is accompanied with apraxia.

Movement disorders: Parkinson's disease, Huntington's disease, olivopontocerebellar atrophy, progressive supranuclear palsy, and Wilson's disease may cause dysphagia by involvement of the extra-pyramidal pathway.

Multiple sclerosis: This autoimmune disorder causes involvement of multiple white matter tracts. Other features include bilateral INO, heat sensitivity, fatigue, sensory impairment.

Amyotrophic lateral sclerosis: It includes spinal and bulbar presentation. The bulbar presentation includes dysarthria, dysphagia, dysphonia, sialorrhoea, muscle atrophy, and fasciculations.

REFERENCES

1. Heemskerk A-W, Roos RAC. Aspiration pneumonia and death in Huntington's disease. *PLoS Curr.* 2012;4:RRN1293. doi:10.1371/currents.RRN1293.
2. Barer DH. The natural history and functional consequences of dysphagia after hemispheric stroke. *J Neurol Neurosurg Psychiatry.* 1989;52(2):236–241.
3. Lazarus C, Logemann AJ. Swallowing disorders in closed head trauma patients. *Arch Phys Med Rehabil.* 1987;68(2):79–84.
4. Kalf JG, De Swart BJM, Bloem BR, Munneke M. Prevalence of oropharyngeal dysphagia in Parkinson's disease: a meta-analysis. *Parkinsonism Relat Disord.* 2012;18(4):311–315.
5. Bray BD, Smith CJ, Cloud GC, et al. The association between delays in screening for and assessing dysphagia after acute stroke, and the risk of stroke-associated pneumonia. *J Neurol Neurosurg and Psychiatry.* 2016;88(1):25 LP-30. http://jnnp.bmj.com/content/88/1/25.abstract.
6. Heemskerk A-W. H1Dysphagia in Huntington's disease: symptoms and a patient perspective. *J Neurol Neurosurg and Psychiatry.* 2016;87(Suppl 1):A56 LP-A57. http://jnnp.bmj.com/content/87/Suppl_1/A56.4.abstract.
7. Daniels SK, McAdam CP, Brailey K, Foundas AL. Clinical assessment of swallowing and prediction of dysphagia severity. *Am J Speech-Language Pathol.* 1997;6(4):17–24.
8. Daniels SK. Neurological disorders affecting oral, pharyngeal swallowing. *GI Motil online.* 2006.
9. Daniels SK, Brailey K, Priestly DH, Herrington LR, Weisberg LA, Foundas AL. Aspiration in patients with acute stroke. *Arch Phys Med Rehabil.* 1998;79(1):14–19.

18 | Neck Pain

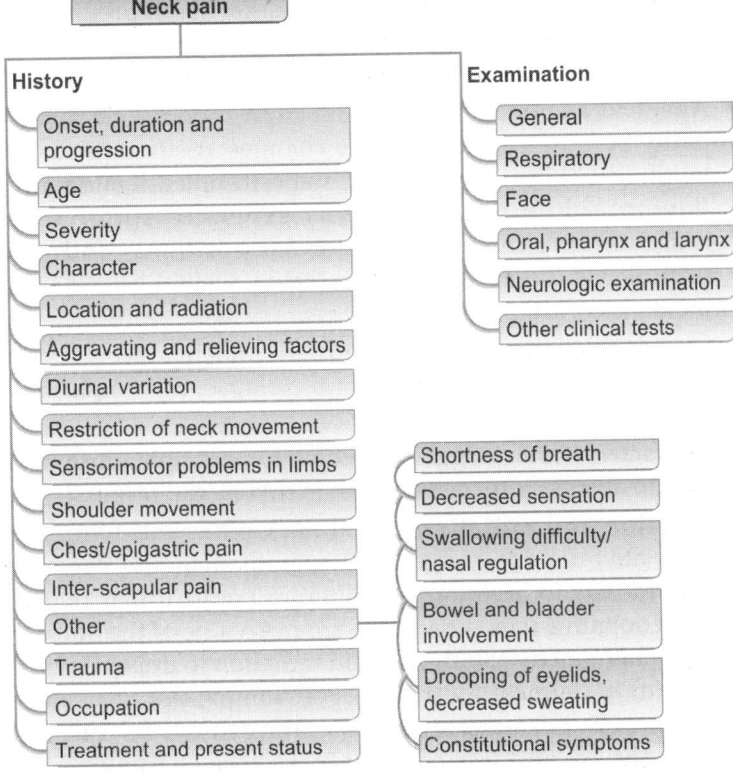

Neck pain

History
- Onset, duration and progression
- Age
- Severity
- Character
- Location and radiation
- Aggravating and relieving factors
- Diurnal variation
- Restriction of neck movement
- Sensorimotor problems in limbs
- Shoulder movement
- Chest/epigastric pain
- Inter-scapular pain
- Other
- Trauma
- Occupation
- Treatment and present status

Examination
- General
- Respiratory
- Face
- Oral, pharynx and larynx
- Neurologic examination
- Other clinical tests

- Shortness of breath
- Decreased sensation
- Swallowing difficulty/nasal regulation
- Bowel and bladder involvement
- Drooping of eyelids, decreased sweating
- Constitutional symptoms

INTRODUCTION

The most widely cited study by Radhakrishnan et al shows the incidence of cervical radiculopathy (CR) to be 107.3/100,000 in men and 63.5/100,000 in women.[1] As a neurologist or neurosurgeon, we have to differentiate first between a cervicogenic or non-cervicogenic pain, and if it is cervicogenic then to find out the cause behind it.

History

- **Onset, duration and progression:** One should ask whether it is acute or chronic. Causes of acute onset neck pain disc prolapse, traumatic fractures, infections and inflammations, and malignancies. Chronic neck pain can be because of diskogenic, or cervical spondylotic myelopathy—patient may present exacerbating and relieving symptoms. CR onset may be acute in half, subacute in quarter, and insidious in quarter. One should enquire about progression —whether it is progressive or non-progressive. Non-progressive suggests myofascial cause of pain.

- **Age:** Old age patient may suffer from spondylotic myelo-pathy because of degenerative changes, ossified posterior longitudinal ligament (OPLL), hypertrophied ligamentum flavum; while young patients generally suffer from muscular spasm (due to abnormal positions) and disc protrusions (because of trauma, heavy weight lifting, etc.). Incidence of CR peaks at 40–50 years of age.[1]

- **Severity:** Severe pain may disturb the patient's sleep. It may occur in disk prolapse, vertebral collapse, inflammations, neoplasms.

- **Character:** It may suggest etiology, for example, throbbing pain in abscess, dull aching pain in myofascial origin.

- **Location and radiation:** Axial neck pain, and whiplash associated disorder, is generally paramedian and vague. It may radiate to occipital, shoulder, or parascapular region. Radicuopathic pain, however, radiates to one of the limbs in a dermatomal distribution.[2] Pain radiation to upper limbs is helpful in delineating the level of compression. It is most commonly because of nerve impingement by disc, or osteophtyes, and less commonly because of trigger point from myofascial pain.

- **Aggravating and relieving factors:** Cervical disc pain, with root involvement, increases with head bending towards involved side, gets relieved with arm elevation. Similarly, activities with increased loading on disc like lifting, Valsalva manoeuvre increases the pain, and lying supine decreases it.

- **Diurnal variation:** Hand paraesthesias increase in night in carpal tunnel syndrome. Pain is also worse in night in subacromial bursitis and frozen shoulder. Morning stiffness that improves over the day, may be linked to rheumatic causes.

- **Restriction of neck movement:** It suggests extradural CVJ anomaly (mostly bony-like AAD). Clicking sounds, neck tilt are also suggestive of CVJ anomaly.

- **Sensorimotor problems in limbs:** In a long standing case of CR patient may present with numbness, paraesthesias or weakness of the corresponding myotome. Table 18.1 describes the distribution of sensory and motor deficits associated with CR.

- **Shoulder movement:** Restriction of shoulder movement is suggestive of local shoulder pathology and not because of radicular cause. Pain on raising the arm up to 90 degrees and then lowering down suggests subacromial bursitis. In this, pain does not occur on raising the arm above 90 degrees.

- **Chest/epigastric pain:** These pains may radiate to upper limbs. Radiating pain is more diffuse as compared to radicular pain which occurs specifically along the dermatome.

- **Interscapular pain:**[3] Along with the radicular pain to the upper limb, Cloward[1] described that pain from the C6-C7 disc was referred to the inferior angle of the scapula. Pain from the C5-C6 disc was referred to the centre of the medial scapular border. Pain from C4-C5 disc was experienced in the region of the spine and superior angle, and pain from the C3-C4 disc was referred to the C7 spinous process and the posterior border of the trapezius muscle.

Table 18.1: Sensory and motor deficits associated with CR[5]			
Level	*Sensory deficit*	*Motor deficit*	*Reflex involved*
C5	Outer aspect of arm	Deltoid	Biceps
C6	Outer aspect of forearm, radial two digits	Biceps, wrist extension	Brachioradialis
C7	Middle finger	Triceps, wrist flexion	Triceps
C8	Ulnar two digits	Finger flexors	Finger flexion
T1	Inner aspect of forearm	Hand intrinsics	

- **Other symptoms:** Shortness of breath: Phrenic nerve involvement, medullary compression. Decreased sensation over face suggests spinal tract of trigeminal nerve involvement. Swallowing difficulty/nasal regurgitation: Lower cranial nerve involvement in CVJ lesions. Bowel and bladder involvement: Due to autonomic involvement. Drooping of eyelids, decreased sweating: Horner's syndrome. Constitutional symptoms: Fever, weight loss, cough, hemoptysis suggest tuberculosis/malignancy.
- **Trauma:** History of recent and past trauma should be taken. Though the incidence of CR preceded by trauma is low.[1,4]
- **Occupation:** Risk factors may be related to occupation, and it has been found that CR is more common in people who lift heavy weights, driving vehicles which vibrate, frequent diving from board, and golfers.[4] On the other hand, myofascial pain is more common in people who place their neck in wrong postures for prolonged duration like professionals working on computers, students, etc.
- **Treatment taken and present status** should be asked. History of improvement or not improvement with any drug like antitubercular drugs, or improvement with application with cervical collar (suggests extradural lesion).

Examination

1. **Posture:** Patients may present with poor posture. Myofascial pain patients present with rounded shoulders and protracted scapulae.
2. **Trigger points:** These are frequently noted in the trapezius, supraspinatus, infraspinatus, rhomboids, and levator scapulae muscles in myofascial pain.
3. **Palpable taut bands:** Palpation of soft tissues, bony, and other structures like lymph nodes, thyroid, etc., should be performed. Palpable taut bands are present in myofascial pain. The palpable, taut band is noted in the skeletal muscle or surrounding fascia; a local twitch response often can be reproduced with palpation of the area.
4. **Range of movement:** Cervical full range of motion including flexion, extension, lateral bending, and rotation should be noted, along with shoulder movements. If myofascial pain is present, then the pain is reproduced in positions that stretch the affected muscle. For example, pain

during leaning away from the affected side incites pain in myofascial origin, while if pain occurs leaning towards the affected side it means it is of cervical radiculopathy.

5. **Sensory examination:** In myofascial pain, axial spondylotic pain, sensory examination will be normal. While in pain due to CR, there will be a degree of sensory loss in the affected dermatomal distribution.

6. **Motor examination:** In myofascial pain, motor examination will be normal. In axial spondylotic pain, again it will be normal. But in CR, the affected myotome group may be having signs of LMN involvement, i.e. may have decreased bulk, tone, power, and deep tendon reflexes. Patients of myelopathy will have signs of UMN involvement, i.e. may have normal or decreased bulk, increased tone, decreased power, and increased deep tendon reflexes.

7. **Manoeuvres:**
   - Pain during head bending towards the affected side occurs in CR.
   - Spurling's sign or manoeuvre: Radiating pain or paraesthesias during head extension and tilting towards the symptomatic side.
   - Cervical distraction test: Relief of pain with manual upward neck traction, with neck in slight flexion.
   - Lhermitte's sign: During flexion of the neck, if an electric shock like sensation runs down the spine, it is indicative of myelopathy.

Investigations

In chronic neck pain, these are ordered either for intractable nature of more than 6–8 weeks (no clear guidelines available), or if there is a red flag sign like a neural deficit. In a traumatic case there are guidelines available including Canadian C-spine rule, NEXUS decision instruments, and American College of Radiology appropriateness.[6–8]

1. **X-rays:** One should look for disc alignment along anterior vertebral body (VB) line, posterior VB line, laminar line, and spinal line. VB and disc height should be noted. Integrity of pars should also be noted. Osteophytic spurs, ossified posterior longitudinal ligament, ossified ligamentum flavum, should be looked for. Apart from AP and lateral

views, oblique views can be added for visualising foramen better. Hard disc herniation can also be seen with it.

2. **CT cervical Spine:** Suspected details on X-rays can be better delineated on CT. It can be combined with angiogram if surgery is planned.

3. **MRI cervical Spine:** Soft disc herniation, signs of myelo-malacia can be seen.

4. **CT myelogram:** It is used only in cases, where MRI is contraindicated.

5. **EMG study:** It is useful for differentiating peripheral nerve entrapment syndromes (like carpal tunnel syndrome) from CR. But the results should be interpreted along with the imaging findings, as this test is associated with high false positive and negative findings.[9,10]

6. **Diskography:** It is a functional test where the contrast is injected in the disc under fluoroscopic guidance, and symptoms are reproduced if the source of the pain is disc.

7. **Rheumatoid factor:** It is ordered when rheumatoid arthritis is suspected.

Differential Diagnosis

- **Myofascial pain:** This is the most common cause, and particularly present in young population having wrong postures, sleep habits, or stress.
- **Cervical radiculopathy**: CR may be because of diskogenic origin, facet joint disease, osteophytic spurs which narrow the foramen and cause compression or destruction of the root. Most common root to be compressed is C7 root by C6-C7 disc. It is accompanied by sensory and motor involvement in the arms and neck.
- **Cervical spondylosis:** The causes include diskogenic pain, facet joint disease, vertebral body disease. There will be history of chronic neck in an old patient, with localised tenderness of the cervical spine, decreased range of motion, and no signs of neural compromise.
- **Cervical myelopathy**: Cord compression due to either soft/ hard disc, OPLL, ossified ligamentum flavum, tumour, vascular malformations, or due to ischemia may lead to myelopathy. It is accompanied with long tract signs.
- **Whiplash associated disorder:** There will be history of motor vehicle accidents, collision, or acceleration–

deceleration forces to the neck. There are 4 grades defined by Quebec classification:[11]

- **Grade I** includes non-specific complaints like pain, stiffness in the neck without objective physical findings.
- **Grade II** includes neck pain with signs limited to musculo-ligamentous structures.
- **Grade III** includes neck pain with neural compromise.
- **Grade IV** includes neck pain with fracture/dislocation.

- **Rheumatoid arthritis:** There may be known history of rheumatoid arthritis in the patient. Radiation may be there up to the back of head. There may be sensory loss in the hands and feet. Examination may range from only cervical tenderness to a myelopathic picture. There may be rheumatoid nodules over the extensor aspect of hands, wrists, and elbows.

- **Spasmodic torticollis:** It may be congenital, irritative/destructive neural lesion, or post-traumatic. Examination will reveal unilateral contracted neck muscles tilted down towards, and rotated away from the involved side. Tenderness may be present. No signs of neural compromise will be present.

- **Cervical fracture and dislocations:** There will be history of trauma, with accompanying sensorimotor deficits.

- **Malignancy:** It may be primary or secondary (metastatic). There will be history of localised pain (worsening at night), fatigue, and weight loss. Examination findings may or may not reveal signs of neurological compromise. In secondary malignancies there may be history of primaries at other places like breast, lungs, genitourinary, prostate, etc. Lymphadenopathy or hepatosplenomegaly may be present. Examination findings are same as of primary malignancies.

- **Vertebral osteomyelitis:** Patient may have low grade fever, local pain, with worsening at night. Examination will reveal tender points, and neural compromise. MRI will show T1 hypointensity, T2 hyperintensity, with contrast uptake. ESR and CRP will be elevated. Abscess may also ensue.

- **Meningitis:** History may be headache, stiffness of neck, fever, altered mental status, with a short duration. Examination findings will be nuchal rigidity, positive Brudzinski's and/or Kernig's sign.

- **Shoulder problems:** Tendinitis, bursitis, impingement syndrome may occur. Symptoms will be more localised to the shoulders, with restriction of its range of motion.
- **Fibromyalgia:** It is a diagnosis of exclusion. Patient may complain of multiple joint pains.
- **Referred pain** to the neck, arm or shoulder can result from the heart, lungs, oesophagus, or upper abdomen.

REFERENCES

1. Radhakrishnan K, Litchy WJ, O'Fallon WM, Kurland LT. Epidemiology of cervical radiculopathy. A population-based study from Rochester, Minnesota, 1976 through 1990. *Brain*. 1994;117 (Pt 2): 325–335.

2. Rhee JM, Yoon T, Riew KD. Cervical radiculopathy. *J Am Acad Orthop Surg*. 2007;15(8):486–494.

3. Cloward RB. Cervical diskography. A contribution to the etiology and mechanism of neck, shoulder and arm pain. *Ann Surg*. 1959;150:1052–1064.

4. Kelsey JL, Githens PB, Walter SD, et al. An epidemiological study of acute prolapsed cervical intervertebral disc. *J Bone Joint Surg Am*. 1984;66(6):907–914.

5. Iyer S, Kim HJ. Cervical radiculopathy. *Curr Rev Musculoskelet Med*. 2016;9(3):272–280. doi:10.1007/s12178-016-9349-4.

6. Stiell IG, Wells GA, Vandemheen KL, et al. The Canadian C-spine rule for radiography in alert and stable trauma patients. *JAMA*. 2001;286(15):1841–1848.

7. Hoffman JR, Mower WR, Wolfson AB, Todd KH, Zucker MI. Validity of a set of clinical criteria to rule out injury to the cervical spine in patients with blunt trauma. National Emergency X-Radiography Utilization Study Group. *N Engl J Med*. 2000; 343(2):94–99. doi:10.1056/NEJM200007133430203.

8. Keats TE, Dalinka MK, Alazraki N, et al. Cervical spine trauma. American College of Radiology. ACR Appropriateness Criteria. *Radiology*. 2000;215:243–246.

9. Alrawi MF, Khalil NM, Mitchell P, Hughes SP. The value of neurophysiological and imaging studies in predicting outcome in the surgical treatment of cervical radiculopathy. *Eur spine J Off Publ Eur Spine Soc Eur Spinal Deform Soc Eur Sect Cerv Spine Res Soc*. 2007;16(4):495–500. doi:10.1007/s00586-006-0189-6.

10. Caridi JM, Pumberger M, Hughes AP. Cervical radiculopathy: a review. *HSS J*. 2011;7(3):265–272. doi:10.1007/s11420-011-9218-z.

11. Suissa S, Harder S, Veilleux M. The Quebec whiplash-associated disorders cohort study. *Spine (Phila Pa 1976)*. 1995;20(8S):12S–20.

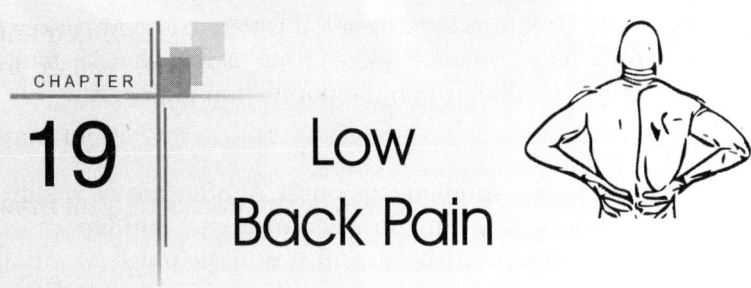

Low Back Pain

Low back pain

History

- Onset
- Duration and progression
- Age and sex
- Occupation
- Radiation
- Relation and position
- Aggravating and relieving factors
- Claudication distance
- Foot drop
- Bowel and bladder disturbance
- Sensory disturbance
- Menstrual disturbances
- Trauma
- Chronic illness
- Treatment history and present status

Examination

- Tenderness
- Deformities
- Gibbus
- Straight leg raising test
- Crossed SLR
- Reverse SLR
- Test for SI joint
- FABER test
- Neurological examination

As a neurologist/neurosurgeon we have to characterize the patients of low back pain into one of the three categories[1]: Chronic non-specific pain, back pain associated with diskogenic/canal stenosis pain, and pain associated with specific spinal cause (serious causes including tumour, infection, cauda equina syndrome). It is very common, and in Indian study its prevalence ranges from as low as 6.2% to as high as 92% depending upon the population under study.[2,3]

History

- **Onset:** It may be spontaneous onset, or after heavy weight lifting, prolonged sitting, or after a trauma. Sudden onset causes are prolapsed disks, and traumatic ones. Gradual onset ones are osteoarthritis, ankylosing spondylitis, tumours, etc.

- **Duration and progression:** Degenerative diskogenic pain is of long duration, traumatic diskogenic pain is of short duration. Pain due to Potts spine is of intermediate duration (of months), while a pain of neoplastic origin may be of long duration (years).

- **Age and Sex:** Diseases of young adults are postural problems, ankylosing spondylitis, trauma; middle aged are: Prolapsed disks, obesity, bursitis; elderly are: Osteoporosis, osteoarthritis, etc. Tuberculosis can present at any age. Females may have gynecological diseases having pain referred to back. Ankylosing spondylitis is commoner in males.

- **Occupation:** Enquire what they have done in last few days. They may give history of heavy weightlifting.

- **Radiation:** Single level radiculopathy is characteristic of diskogenic pain particularly if it is going below knee and in the posterior aspect,[4] while multilevel radiculopathy is characteristic of spinal stenosis.

- **Relation with position:** Discogenic pain is least in lying down position, increases in standing, sitting, and stooping forward position. Canal stenosis pain decreases in bending forwards, walking uphill and riding bicycle because it causes flexion and opens up the canal. Pain of recess stenosis occurs on standing, and is absent in sitting. Postural pain occurs after attaining that posture for a long time, e.g. in people with prolonged sitting; and such pain weans off after activity.

- **Aggravating and relieving factors:** Bending forward increases pressure over the disc and anterior structures of a verterbrae, and thus leads to increase in the symptoms of diskogenic pain; while bending backward causes buckling of ligamentum flavum into the spinal canal, thus increasing the symptoms of spinal stenosis. Pain of canal stenosis can increase with exercise and prolonged standing (neurogenic claudication), and get relieved with change in posture and rest. Pain of radiculopathy can increase with coughing, straining, sneezing. Standing on one leg increases pain in sacroiliac disorders.

- **Claudication distance:** It decreases with increasing severity.
- **Foot drop:** It may occur if L5 root is compressed which leads to the motor complaints.
- **Bowel and bladder disturbance:** Fecal incontinence/ urinary retention may be due to compression at cauda equina roots. Hesitation or incomplete micturition may occur due to compression on cord from outside (delayed presentation), or from within (intramedullary—early presentation).
- **Sensory disturbance:** Sensory loss may occur either after the radicular pain when permanent root injury has occurred, or without the history of radicular pain when cord compression or lesion is there.
- **Menstrual disturbance:** It should be asked from every female as gynecological problems like retroverted uterus may have pain referred to back.
- **Trauma:** It should always be enquired.
- **Chronic illness:** Past significant history including malignancy, IV drug abuser, history of appetite and weight loss points (tuberculosis, malignancy) should be enquired. Fever may point towards tuberculosis. Diabetes mellitus may be associated with autonomic and sensory symptoms. Osteoporosis/old age may be associated with spontaneous fractures or listhesis.
- **Treatment history and present status** of the pain should also be enquired. Treatment history of antituberculosis treatment with improvement indirectly confirms the diagnosis of Pott's spine. Drug treatment with steroids and phenytoin leading to osteoporosis or hypocalcemia increases the risk of fractures.

Examination

- **Tenderness:** Midline tenderness suggestive of Pott's malignancy, fracture. While paramedian tenderness is suggestive of facet joint disease. Tenderness at costovertebral angle suggests renal, adrenals or injury to transverse process.

- **Deformities:** Scoliosis should be looked for. Loss of lumbar lordosis may occur due to spasm.
- **Gibbus:** Suggestive of Pott's disease.
- **Straight leg raising test (Lasègue test):** Pain on flexing the hip (with the knee extended). It is the main test which detects radicular compression of L5 and S1 roots. Pain occurs between 30° and 70°. Pain below 30° suggests non-organicity, while pain above 70° can be in normal persons also. SLR is absent in recess stenosis.
- **Crossed SLR:** Pain on raising the opposite (sound) leg increases the specificity.
- **Bragard's sign:** While doing SLR test, patient's foot is dorsiflexed. It increases the sensitivity of the test.
- **Sicard's sign:** While doing SLR test, patient's great toe is dorsiflexed. It further increases the sensitivity.
- **Reverse SLR (Femoral stretch or Ely test):** It is raising the leg in the prone position. It denotes compression of the high lumbar region, i.e. femoral nerve.
- **For sacroiliac disease:** Flex the knee and the flex the hip. If patient experiences pain, it indicates SI disease.
- **FABER (Flexion, ABduction, and External Rotation) test:** It detects the hip joint disease. With the knee bent, the thigh is flexed, abducted and externally rotated. If pain is felt, then the hip joint is diseased.
- **Examination of spine:** Tenderness should be searched for on the spinous process, facet joints (lateral to the midline). Movement of the spine should be checked in all directions, i.e. flexion, extension, and lateral deviation, and pain or restriction in them should be noted.
- **Sensorimotor examination of the lower limbs (see concerned chapters):** Examine the sensory modalities in the lower limbs (all modalities may be lost). Examine the power, particularly, in the dorsiflexors. Foot drop may occur because of L5 root lesion; EHL weakness is characteristic of L5 lesion. Look for atrophy and fasciculations. Reflexes of knee and ankle should be checked which tell about L3–L4, and S1 roots respectively. Medial hamstring reflex tells

about L5 root. Anal wink reflex, Bulbocavernosus reflex tells us about the S2, S3, S4 levels.

Investigations

- **X-rays:** This is generally the initial investigation. Bony changes including destruction, fractures, listhesis, end plate changes, intradiscal vaccum sign may be visible. Suspected listhesis may be confirmed with dynamic (flexion-extension) films.
- **MRI lumbo-sacral spine:** It is the investigation of choice, as it delineates the soft tissue well. Pathologies like herniated discs, inflamed facets, canal stenosis, recess stenosis, synovial cyst, tumours, tuberculosis, hematomas, are well shown. Urgent MRI is indicated in red flag signs which are: Cauda equine syndrome, weakness/foot drop, intractable pain, suspicion of malignancy.
- **CT:** Bony pathologies suspected on X-rays can be confirmed on this.
- **CT myelogram:** Its use has decreased in the present era. Only indication left probably is in a patient with a metallic implant on which MRI cannot be performed.
- **EMG/NCS:** These are indicated for progressive motor weakness. In normal examination these tests are nearly always normal.
- **Serum tests—ESR and CRP** are indicated for suspected chronic inflammatory diseases like tuberculosis, ankylosing spondylitis (AS), sarcoidosis, etc. These are also indicated in the follow-up or for seeing treatment response with drugs. **HLA B-27** is indicated for AS.
- **Chest X-ray:** It is also indicated for suspected tuberculosis, sarcoidosis, suspicion for primary malignancy which has metastasised to spine.
- **USG abdomen:** It is indicated for suspicion of primary malignancy, tuberculosis, pelvic inflammatory disease (in females) (Table 19.1).

Differential diagnoses of low backache

- **Non-specific low back pain:** More than 85% of patients presenting to clinics with low back pain come in this entity.[6] No characteristic findings will be seen.

Table 19.1: Difference between neurogenic and vascular claudication[5]

Manifestation	Neurogenic claudication	Vascular claudication
Location of pain	Back, buttocks, legs; bilateral	Calf; unilateral
Associated symptoms	Paresthesias, weakness, priapism, incontinence	None
Provoking factors	Prolonged standing	Exercise, walking
Relieving factors	Sitting, lying down	Rest
Walking uphill	No pain	Increases the pain
Bicycling	No pain	Increases the pain
Time for relief	Minutes	Seconds
Effect of back hyperextension	Reproduces symptoms	No effect
Pulses	Normal	Decreased
Neurologic exam after exercise	Weakness, loss of reflexes	No change

- **Herniated nucleosus pulposus:** Low backache, referred and radicular pain to the distribution of the root (radicular), pain decreased by lying down, SLR is positive, myotomal weakness (according to supply), reflexes decreased (according to the roots), age 30–55 years.
- **Lateral recess stenosis:** Low backache, referred and radicular pain according to the root involved, pain increases on standing and exertion, and decreases in lying position; SLR is negative; muscle weakness may occur.
- **Discogenic pain:** Low backache, referred pain to buttock, posterior thigh, no radicular pain; pain increases on sitting and flexion, and relieves on lying and extension position; SLR negative; motor symptoms absent.
- **Facet pain:** Low backache; referred pain to buttock, posterior thigh; no radicular pain; pain increases on lying supine, extension, and rotation; relieves on flexion; SLR negative; motor symptoms and signs absent.
- **Musculo-ligamentous pain:** Low backache; referred pain to buttock, posterior thigh; no radicular pain; pain increases on bending, minor movements; relieves on rest, lying; SLR negative; motor symptoms or signs absent.

- **Spinal stenosis:** Low backache, multilevel radiculopathy, cauda equina syndrome; referred pain to buttock, posterior thigh; pain increases on exercise, standing; SLR variable; motor symptoms according to root involved.
- **Pott's spine:** Pain predominant over midline; tenderness present; knuckle (single level), gibbus (up to 3 levels), or kyphosis (more than 3 levels) present depending on the number of levels involved; SLR positive; UMN signs may be present if cord involved. Constitutional symptoms like fever, fatigue, weight loss may be present.
- **Hip joint disease:** Pain predominant in hips and buttocks; referred pain to groin, anterior thigh, lateral thigh, knee; no radicular pain; SLR negative; FABER test positive; no motor symptoms or signs.
- **Tumours:** They may be primary or secondary (metastatic). Primary tumours in this region include myxopapillary ependymoma, nerve sheath tumours (schwannoma), meningioma, arachnoid cyst. Presentation may be gradually progressive pain localised to the region, more in reclining and in night, aggravated by exertion and straining. Pain, sensory disturbance and motor weakness may be present depending upon the structures involved.
- **Metastatic/osteomyelitic pain:** Low backache; referred pain to buttocks and thighs; no radicular pain; no relation to position; SLR negative; no motor symptoms and signs.
- **Viscerogenic pain:** Variable localization; variable referred pain; SLR negative; no motor symptoms or signs. For example, pain from retroperitoneal diseases.

REFERENCES

1. Chou R, Qaseem A, Snow V, et al. Diagnosis and treatment of low back pain: a joint clinical practice guideline from the American College of Physicians and the American Pain Society. *Ann Intern Med.* 2007;147(7):478–491.
2. Ahdhi GS, Subramanian R, Saya GK, Yamuna TV. Prevalence of low back pain and its relation to quality of life and disability among women in rural area of Puducherry, India. *Indian J Pain.* 2016; 30(2):111.
3. Bindra S, Sinha AKG, Benjamin AI. Epidemiology of low back pain in Indian population: A review. *Int J Basic Appl Med Sci.* 2015;5(1): 166–179.

4. Vroomen P, De Krom M, Knottnerus JA. Diagnostic value of history and physical examination in patients suspected of sciatica due to disc herniation: a systematic review. *J Neurol*. 1999;246(10):899–906.

5. Campbell WW. *DeJong's THE Neurologic Examination*. Seventh. Philadelphia: Lippincott Williams and Wilkins; 2013.

6. van Tulder MW, Assendelft WJJ, Koes BW, Bouter LM. Spinal radiographic findings and nonspecific low back pain: a systematic review of observational studies. *Spine (Phila Pa 1976)*. 1997;22(4): 427–434.

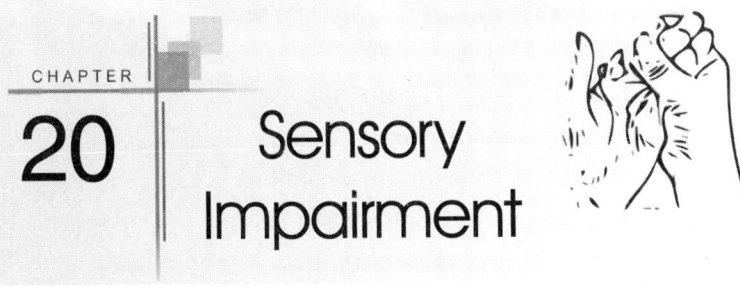

CHAPTER
20
Sensory Impairment

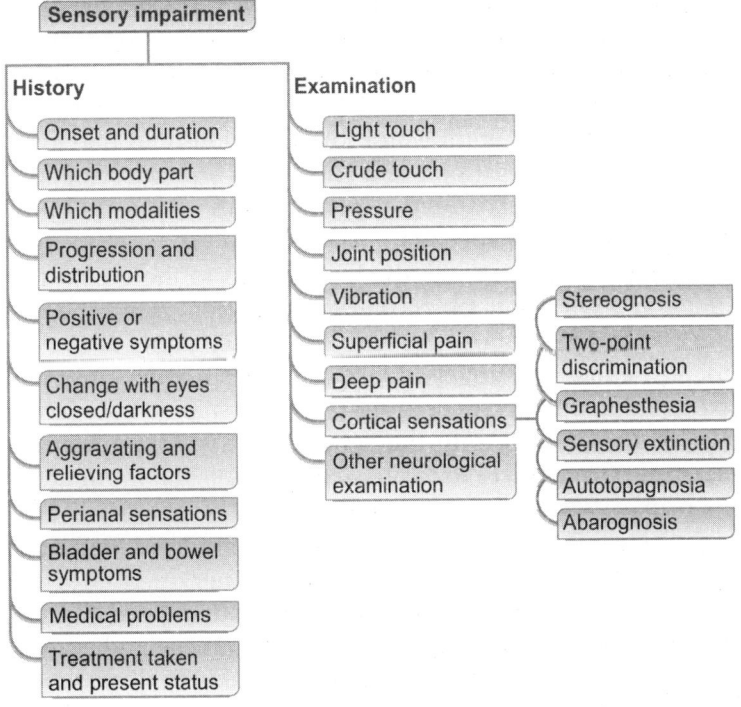

Sensory impairment

History
- Onset and duration
- Which body part
- Which modalities
- Progression and distribution
- Positive or negative symptoms
- Change with eyes closed/darkness
- Aggravating and relieving factors
- Perianal sensations
- Bladder and bowel symptoms
- Medical problems
- Treatment taken and present status

Examination
- Light touch
- Crude touch
- Pressure
- Joint position
- Vibration
- Superficial pain
- Deep pain
- Cortical sensations
- Other neurological examination
  - Stereognosis
  - Two-point discrimination
  - Graphesthesia
  - Sensory extinction
  - Autotopagnosia
  - Abarognosis

INTRODUCTION

It is rare to have a presentation with pure sensory loss. Generally, patient presents with combined sensory and motor involvement. We will discuss, however, all differentials for sensory loss.

Accepted Definitions

- Algesia: Sense of pain
- Hypalgesia: Decrease in pain intensity
- Analgesia: Absence of pain
- Hypesthesia: Decrease in the sensations
- Anaesthesia: Absence of all sensations
- Paraesthesia: Abnormal sensations
- Dysesthesia: Abnormal unpleasant or painful sensations

Anatomy of the sensory system: Detection of the stimulus is done by receptors present in skin, muscles and tendons. These are free nerve endings for pain and temperature, pacinian corpuscles for vibration and pressure, Meissner's corpuscles for fine touch (abundant in thick hairless skin like fingerpads), Merkel's cell nerve endings for low vibrations and fine touch (abundant in finger tips), and Ruffini's corpuscles for joint position (located in deep layers of skin and sense joint deformation with a specificity of 2.75 degrees).

Peripheral nerves carry the impulse up to the dorsal root ganglion, from where the second order neuron starts. The pain and temperature fibres get crossed 1–2 segments above the point of entry and travel either in lateral spinothalamic tract (pain and temperature) or anterior spinothalamic tract (crude touch). These synapse in the ventroposterolateral nucleus of thalamus, from where the third neuron starts and terminate in the primary somatosensory cortex.

The fine touch, vibration and pressure sensations are carried up to the dorsal root ganglia by peripheral nerves, from where the second order neuron starts and travel uncrossed through fasciculus gracilis and cuneatus up to the gracile and cuneate nucleus in the medulla. Thereafter these fibres decussate as internal arcuate fibres and form medial leminiscus, and then relay in the primary somatosensory cortex.

History

- **Onset and duration:** Acute onset points towards a vascular cause, e.g. Wallenberg syndrome (PICA territory infarct), insidious onset suggests infectious or inflammatory cause, e.g. meningitis or meningoencephalits, while chronic ones include neoplasms or degenerative disorders.

- **Which body part:** Ask in which body part, patient is experiencing sensory loss.
- **Which modalities:** Ask for which modalities, pain, temperature, pressure, etc. patient is experiencing loss of sensations.
- **Progression and distribution:** Whether the condition is progressive or non-progressive. Ask whether it is rapidly progressive. Enquire how the sensory loss progressed: Which limb was involved first, and how the other limbs got involved. Sometimes pain in a region is radicular in nature and was localized somewhere else, e.g. diskogenic pain. Distal to proximal progression of sensory impairment occurs in neuropathy/axonopathy; proximal to distal is rare/non-neurologic. Progression first in upper limbs, then lower limbs, i.e. descending type occurs in intramedullary lesions, while reverse, i.e. ascending occurs in extramedullary lesions.
- **Positive or negative symptoms:** Patient may either feel positive symptoms like pins and needle sensations, pain, hyperpathia, paraesthesias like crawling insects, burning etc., or he may be feeling negative symptoms like numbness, anaesthesia. Ask whether he can feel pin pricks, needle pricks, thorn at ground, pinching (for pain), touch of clothes (for touch), hot and cold water while taking bath and holding such different temperature utensils (for temperature). Positive symptoms like paraesthesias have localizing value.
  Ask whether he feels heaviness of limbs (numbness), tight band like sensations over any part of body. History of slippage of chappals without knowledge (posterior column involvement or peripheral nerve involvement), history of painless burns while cooking suggests dissociated sensory loss (syringomyelia); history of difficulty in identifying common things like coins, keys, pencils (astereognosis).
- **Change with eyes closed/darkness:** Increase in the symptomatology in the darkness or with eyes closed indicates proprioceptive impairment, or a vestibular pathology. Patients feel ataxia in the night and in darkness.
- **Aggravating and relieving factors:** If the condition is aggravated by walking, coughing, sneezing and straining, then it may be a radicular pathology. If it aggravates by neck movement, it is a Lhermitte's sign.

- **Perianal sensations:** Loss of perianal symptoms are a characteristic symptom of cauda equina syndrome.
- **Bladder and bowel symptoms:** These may be lost early in intramedullary lesions, and are lost late in extramedullary lesions.
- **Medical problems:** Diabetes, thyroid and other medical problems should be enquired as these may be the cause of peripheral neuropathy.
- **Treatment taken and present status:** Chemotherapy, like cisplatin, carboplatin can cause sensory disturbances.

Examination: Go from abnormal to normal if complaint is anaesthesia, and reverse if complaint is hyperalgesia.

- **Superficial pain:** Test with a "safety pin", or an "All pin". Do not use needles kept in ward.
- **Deep pain:** Press muscle belly (achilles), supraorbital compression (supraorbital nerve).
- **Temperature:** Test with test tubes containing cold and hot water of temperatures 5–10°C and 40–45°C respectively. Temperatures above and below this elicits pain.
- **Crude touch:** Test with finger rub, or brush.
- **Light touch:** Test with a wisp of cotton.
- **Pressure:** It is tested by a firm touch, or by pressure on muscles, tendons and nerves.
- **Joint position and proprioception**: Start testing at meta-tarsophalangeal joint of the great toe. If this is normal, then going proximally is not necessary, otherwise progress in a sequential manner to ankle, knee, wrist and elbow.
- **Vibration:** Use a tuning fork of 128 Hz frequency. Test over bony prominences, i.e. over great toe, malleoli, tibial tuberosity, anterior superior iliac spines, spines of vertebral column, acromion process, sternum, mandible. If distally it is normal, then no need to test it proximally. Difference in time of vibration of 3–5 seconds from opposite side is abnormal.
- **Test Romberg's sign** (one of the earliest sign): Positive in posterior column involvement, and vestibular involvement.
- **Cerebral sensory functions:** These are tested only when primary sensations are intact.

- Stereognosis: Perception, identification and recognition of the object by touch. It is the earliest sign of parietal lobe dysfunction if primary sensations are intact.
- Graphaesthesia: It is the ability to recognize numbers or letters written over hand with a pencil/dull pin.
- Two-point discrimination: Ability to identify two points. May be static and moving. It is normally 1 mm on the tip of tongue, 2–3 mm on lips, 2–4 mm on fingertips, 4–6 mm on the dorsum of fingers, 8–12 mm on the palm, 20–30 mm on the back of hand, 30–40 mm on the dorsum of foot.
- Sensory extinction: Ability to perceive two simultaneous sensory stimuli.
- Autotopagnosia: Inability to identify body parts, orient the body, a defect in the body scheme, e.g. finger agnosia.
- Abarognosis: Inability to differentiate weights.
• Other neurological examination: Examine motor system, cranial nerves to help in localization.

Investigations

1. **Nerve conduction studies and electromyography:** These are helpful in characterizing neuropathies. Lesions proximal or distal to dorsal root ganglia can also be delineated.
2. **MRI brain/spine:** These help in finding out the lesions at their respective parts.
3. **Lab studies:** Serum investigations can be sent for diabetes, HIV, lymphoproliferative diseases, uremia, vitamin B_{12} deficiency, syphilis, Lyme disease, etc. in view of history and clinical examination.

Important Localizing Points

• Romberg's test is earliest sign of Posterior column disease
• Sensory ataxia is encountered in posterior column disease, severe peripheral neuropathy, dorsal root gangliopathy, vitamin B_{12} deficiency, tabes dorsalis (it is now less seen).
• In subacute combined degeneration, vibration loss can be worse than position sense loss. Reverse is true for tabes dorsalis.
• In parietal lobe lesion: Pain, temperature sensations are preserved, as these get sensed at VPL nucleus of thalamus. Position and vibration are affected (position sense the most).

- In demyelination vibration is often lost first.
- Gradual loss of sensation from toe to ankle favours a peripheral nerve problem, while uniform loss of sensation beyond a certain point, like waist, favours a myelopathy.
- Gerstmann's syndrome: Found in dominant parietal lobe lesions. Components are: Agraphia, finger agnosia, dyscalculia, right-left disorientation.
- Dissociative sensory loss: A loss in which one modality is lost more than other, e.g. in lateral medullary stroke (Wallenberg's syndrome) pain and temperature are lost and posterior column sensations are spared, since spinothalamic fibres lie laterally, and medial leminiscus lies medially. Another example is syringomyelia in which there is pain and temperature loss due to their crossing of fibres in the anterior white column, while posterior column travelling far and posterior gets spared. Anterior spinal artery syndrome is another example in which there is involvement of motor deficits, pain and temperature loss, but preservation of touch, pressure, position and vibration. CNS disease can also lead to dissociation, particularly when there is marked dissociation affecting one body region.
- A lesion of peripheral nerve causes loss of all modalities.

Table 20.1: Site and characteristics of sensory disturbance[1,2]	
Site	Characteristic
Cortex	• Deficits in a hemi distribution
	• Upper limbs are involved more common than lower limbs (because of the arrangement of sensory homunculus)
	• Marked dissociation affecting one body region
	• Loss of cortical sensations
	• Gerstmann's syndrome in left-sided lesions
	• Left-sided lesions cause agnosias on both sides of body, while right-sided lesions causes agnosias on the left side.
Internal capsule	• Dense loss of sensations on the opposite side
	• No pain (when compared to thalamic lesion)
Thalamus	• Impairment of all sensory modalities on the opposite side. These are commonly because of lacunar infarcts,[3] though tumours or abscesses may also result it. Its

Contd.

Table 20.1: Site and characteristics of sensory disturbance[1,2] (Contd.)	
Site	*Characteristic*
	recovery may lead to thalamic pain syndrome, with normal sensory examination, in the contralateral hemibody.[4]
Brainstem	• Crossed deficits, face on one side and body on the opposite side, e.g. Wallenberg syndrome[5]
Spinal cord	• Deficits involving both sides of body below a certain level
	• Different patterns may occur according to lesion, e.g. Transverse syndrome, central cord syndrome, posterior column syndrome, Brown-Séquard syndrome, anterior cord syndrome or conus medullaris syndrome
	• Band-like sensation occurs at the level of lesion
	• Suspended dissociated sensory loss occurs in syringomyelia
	• Lhermitte's sign: Shock-like sensation radiating down the spine occurs in cervical cord lesions, multiple sclerosis, or degenerative lesions. It is because of posterior column involvement. Reverse radiation, i.e. from below upwards occurs in posterior column involvement of lumbar lesion.
	• Sacral sparing suggests intraparenchymal pathology
Dorsal root ganglia	• Subacute onset of pain, paraesthesias and sensory loss which affects large fiber more than small fibers.
	• Strength is preserved, reflexes disappear.
	• Sensory ataxia
Nerve root	• Segmental distribution
	• Decreased sensation
	• Pain, paraesthesias occurs
	• Increased with movement, coughing or straining
	• Root compression signs present, e.g. straight leg raising test
	• Weakness as well as loss of reflex occurs.
Peripheral nerve	• Stocking glove distribution
	• Vibration is often lost first (in generalized forms)
	• In large fibre neuropathies there occurs reflex loss, and when severe there is motor involvement
	• In small fibre neuropathies there occurs burning pain with no motor loss and preserved reflexes.

REFERENCES

1. Campbell WW. *Dejong's the neurologic examination* 7th ed: Lippincott Williams and Wilkins 2013.

2. Gami N. *Bedside approach to Clinical Neurology.* 2nd ed: Jaypee.

3. Kim JS. Pure sensory stroke. Clinical-radiological correlates of 21 cases. *Stroke.* 1992;23(7):983–987.

4. Henry JL, Lalloo C, Yashpal K. Central poststroke pain: an abstruse outcome. *Pain Research and Management.* 2008;13(1):41–49.

5. Kim JS, Lee JH, Lee MC. Patterns of sensory dysfunction in lateral medullary infarction Clinical-MRI correlation. *Neurology.* 1997;49(6):1557–1563.

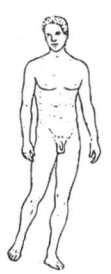

```
                    ┌─────────────────────┐
                    │     Weakness        │
                    └─────────────────────┘
```

History   **Examination**

- Onset and duration   - General examination   - Bulk
- Progression   - Higher mental function   - Tone
- Which limb   - Cranial nerves   - Power
- Spatial progression   - Motor examination   - Reflexes
- Stiffness/   - Sensory examination   - Abnormal
  spasticity   - Cerebellar examination   movements
- Flaccidity   - Coordination
- Endurance   - Gait
- Breathing difficulty
- Bowel/bladder
  involvement
- Sensory disturbances   - Facial weakness
- Associated symptoms   - Speech difficulty
- Depression and   - Visual difficulty
  psychiatric illness   - Chewing difficulty
- Trauma   - Swallowing
- Past history   difficulties/regurgitation
- Treatment taken   - Cervical or backache
  and present status
- Personal history

A motor deficit can be because of cortical, spinal, peripheral nerve related, neuro-muscular junctional, or muscular disorder. Through history and examination, we need to make a provisional diagnosis, and order appropriate investigations to localise it. Etiological diagnosis can also be reached with these.

History

- **Onset and duration:** A sudden onset weakness points towards a vascular cause (stroke, ICH), tumour bleed, disc herniation, trauma; while an insidious onset may point towards a neurogenic cause like tumour expansion.

- **Progression:** Non-progressive etiology may be stroke. Gradually progressive etiology may be a tumour, episodic progressive etiology may be repeated ICHs, tumour bleed, or inflammatory.

- **Which limb:** Ask which limb was involved. Weakness can be classified according to it (monoplegia, diplegia, hemiplegia, quadriplegia).

- **Spatial progression:** Distal to proximal/proximal to distal. Progression from one limb to another limb should be asked.
  - Distal weakness of lower limbs: Patient will present with dragging of feet, difficulty in holding slippers, difficulty in introducing feet in slippers.
  - Proximal weakness of lower limbs: Difficulty in climbing and getting downstairs, difficulty in standing and sitting from squatting position.
  - Distal weakness of upper limbs: Difficulty in buttoning/ unbuttoning, breaking chapattis and difficulty in threading needle (distal muscles of hand), difficulty in rotating door knobs (muscles of wrists), difficulty in bringing glass of water to mouth (muscles of elbow joint).
  - Proximal weakness of upper limbs: difficulty in combing hairs, lifting weight up to head, pouring water over head (muscles of shoulders).
  - Neck muscle weakness: Difficulty in neck holding, raising head above bed

- **Stiffness/Spasticity:** This is an important symptom and sign of upper motor neuron lesion. Mild spasticity is complained as of early fatigue. Lately, he may complain of stiffness while wearing limbs for wearing trousers, frequent falls, not able to bend knees or elbows, flexor spasms, constriction band over abdomen.

- **Flaccidity:** It indicates lower motor neuron lesion
- **Endurance:** It is the ability to perform the same act repeatedly. Lack of endurance is seen in myasthenia gravis, while increase in the performance after a while is seen in Lambert-Eaton syndrome.
- **Breathing difficulty:** It indicates either diaphragmatic weakness, or intercostal muscle weakness, or a phrenic or brainstem lesion.
- **Bowel/bladder involvement** should be asked which may localize the lesion to be of UMN or LMN type (*see* neurogenic bladder).
- **Sensory disturbances:** Absence of sensory disturbance and only motor disturbance may be seen in lesion localized purely to motor area, or a motor neuron disease.
- **Associated symptoms:** History should be taken for headache, seizures, facial weakness, speech difficulty, visual difficulty, chewing difficulty, swallowing difficulty/ regurgitation, fever. These indicate cerebral and cranial nerve involvement. Cervical or backache should also be asked as it may indicate spinal pathology.
- **Depression and other psychiatric illness** should be ruled out in the history, which can cause hysteria or malingering by the patient.
- **Trauma:** Chronic SDH may develop weeks after a trivial trauma, leading to hemiparesis.
- **Past history:** Medical illnesses like hypertension, diabetes, ischemic heart disease (IHD) may precipitate stroke.
- **Treatment taken and present status:** Ask whether patient is currently bed ridden, sitting in bed, walking with support, dependent or independent.
- **Personal history:** History of drug addiction, smoking, alcohol, sexually transmitted diseases should be enquired.

Examination

- **General examination:** One should examine for pulse irregularities, hypertension, cyanosis, clubbing, neurocutaneous markers, lymphadenopathy, shortening of limbs.

- **Higher mental functions (HMF):** Examine HMFs. These are important in ruling out cranial causes.
- **Cranial nerves (CN):** Examine all cranial nerves, specially lower cranial nerves. These may help in localisation of the lesion.
- **Motor examination:**
  - **Bulk:** (By inspection, palpation and measurement)
    - Atrophy occurs in conditions causing lesions of anterior horn cell, nerve root, and muscles. It does not occur in NM junction disorders. Generalized atrophy may occur in UMN lesions if the condition is long standing which leads to disuse atrophy. Other conditions which may cause atrophy are old age, endocrine disorders, and malnutrition.
    - In neurogenic atrophy, the atrophy and weakness are of equal proportion; in myopathy, myasthenia gravis and periodic paralysis the weakness is more than atrophy, while in disuse atrophy, arthritis, endocrine disorders or cachexia, the atrophy is out of proportion to weakness.
    - Arthrogenic atrophy is periarticular.
    - Endocrine disorders which leads to atrophy are hyperthyroidism, hyperparathyroidism, excess cortico-steroids (exogenous or endogenous), hypopituitarism, and diabetes.
    - Congenital hypoplasia or absence of muscle may occur. Prone muscles are depressor angulii oris, palmaris longus, trapezius, peroneus tertius, and anterior abdominal muscles (prune-belly syndrome).
    - In neurogenic atrophy the cause of atrophy is lack of growth cues of anterior horn cells. The slower the pathology progresses, more is the atrophy than the weakness.
    - Single nerve root lesions do not cause much atrophy, because muscles are generally innervated by many nerve roots.
    - Severe cerebral insults in childhood may result in hemiatrophy along with hemiplegia, hemiseizures and hemidystonia (4-hemi syndrome).
    - Hypertrophy of muscles may be physiologic or patho-logic. It is common in myotonia congenita, especially the

dominant form. Androgenic steroids and beta-2 adrenergic agonists may lead to muscle hypertrophy.

- In pseudohypertrophy the muscle fibres get replaced with fat and fibrous tissue. It is common in Duchenne and Becker dystrophy. Calf and infraspinatus muscles are particularly prone.

– **Tone:** By inspection, palpation, and percussion

  - Tests: Babinski-Tonus test, head dropping test, pendulousness of the legs, shoulder shaking test, arm dropping test, hand position.
  - Hypotonicity may develop in lesions of motor unit, cerebellum, proprioceptive pathway (afferent system), parietal lobe, and in chorea. In motor unit disease hypotonicity is accompanied with weakness, while in cerebellar disease weakness is absent. In LMN lesions the hypotonia is severe, while in cerebellar disease the hypotonia is mild. Patients with cerebral stroke initially presents with hypotonia (cerebral or neural shock), which gradually converts into hypertonia.
  - Hypertonicity results from UMN lesions and extrapyramidal lesions. Hypertonicity in CST lesions are more in antigravity group of muscles, while in extrapyramidal lesions it is equal. There is clasp knife spasticity in CST lesions, lead pipe rigidity in extrapyramidal lesions, and cogwheel rigidity in Parkinsonism. Decereberate rigidity occurs when lesion is between red nucleus and vestibular nucleus. Decorticate rigidity occurs when lesion is above red nucleus.
  - Paratonia is alteration in tone to passive motion, and occurs in diffuse frontal lobe disease.

- **Fasciculation:** Can be seen, and palpated. A sign of LMN disease, most common of motor neuron disease; other causes may be radiculopathy, peripheral neuropathy, thyrotoxicosis; myopathies do not cause fasciculations. Hypercaffenism may also cause fasciculations.

- **Power:** It should be graded on Medical Research Council (MRC) scale.[1] It is as follows:

Grade 0: No power

Grade 1: Flicker or trace of contraction

Grade 2: Active movement with gravity eliminated

Grade 3: Active movement against gravity

Grade 4: Active movement against gravity and resistance

Grade 5: Normal power

*Grade 4–, 4, 4+ may be used to indicate movement against slight, moderate, and strong resistance, respectively.

Table 21.1: Movements, their responsible muscles, and their innervations. Major innervations in bold, minor innervations in round brackets, main nerve from which branch has arrived in square brackets

Movements	Muscle	Segmental innervations	Peripheral nerve
Neck			
Flexion	• Sternocleido-mastoid • Platysma • Suprahyoid • Infrahyoid • Scalene (ant med, post) • Longus colli • Rectus capitis	• **CN XI,** C (1), 2, 3 • **CN VII** • V3, VII, C1 • Ansa cervicalis, C1 via CN XII • C4-C7, C4-C8, C6-C8 • C2-C6 • C1-C2	• **CN XI** • **Facial nerve** • Suboccipital nerve
Extension (retraction)	• Trapezius • Paravertebral muscles (Spl enii, erector spinae, trans-versospinalis, and interspinal-intertransverse)	• **CN XI,** C(2), 3, 4 • C1-C8	CN XI
Lateral bending	• Trapezius	• **CN XI,** C(2), 3, 4	
Rotation	• Contralateral sternocleido-mastoid • Ipsilateral trapezius and splenius capitis	• **CN XI,** C(1), 2, 3 • **CN XI,** C(2), 3, 4	
Shoulder/ scapula			
Shrugging /elevation of scapula	• Upper fibres of trapezius	• **CN XI,** C(2), 3, 4	• **CN XI**

Contd.

Table 21.1: Movements, their responsible muscles, and their innervations. Major innervations in bold, minor innervations in round brackets, main nerve from which branch has arrived in square brackets (Contd.)

Movements	Muscle	Segmental innervations	Peripheral nerve
Depression	• Levator scapulae	• C3, C4, C5	• N to Levator scapulae
	• Sternocleidomastoid	• CN XI, C(1), 2, 3	• CN XI
	• Lower trapezius	• CN XI, C(2), 3, 4	• Medial anterior thoracic nerve
	• Pectoralis minor	• C7-T1	• Nerve to subclavius
	• Subclavius	• C5-C6	
Protraction	• Serratus anterior	• C5-C7	• Long thoracic nerve
	• Pectoralis minor	• C7-T1	• Medial anterior thoracic nerve
Retraction	• Rhomboids	• C4-C5	• Dorsal scapular nerve
	• Middle fibres of trapezius	• CN XI, C(2), 3, 4	• CN XI
Flexion	• **Pectoralis major**	• C5-T1	• Medial and lateral pectoral nerve
	• **Anterior fibres of deltoid**	• C5-C6	• Axillary nerve
	• Subscapularis	• C5-C7	• Subscapular nerve
	• Coracobrachialis	• C6-C7	• Musculocutaneous nerve
	• Biceps brachii	• C5-C6	• Musculocutaneous nerve
Extension	• **Posterior fibres of deltoid**	• C5-C6	• Axillary nerve
	• **Latissimus dorsi**	• C6-C8	• Thoracodorsal nerve
	• Triceps	• C6-C8	• Radial nerve
	• Subscapularis	• C5-C7	• Subscapular nerve
	• Teres major	• C5-C7	• Lower subscapular nerve
Adduction	• **Pectoralis major** (sternal portion)	• C5-T1	• Medial and lateral pectoral nerves
	• Latissimus dorsi	• C6-C8	• Thoracodorsal nerve

Contd.

Table 21.1: Movements, their responsible muscles, and their innervations. Major innervations in bold, minor innervations in round brackets, main nerve from which branch has arrived in square brackets (Contd.)

Movements	Muscle	Segmental innervations	Peripheral nerve
Abduction	• Supraspinatus (first 15°)	• C(4), C5, C6	• Suprascapular nerve
	• Deltoid (up to horizontal)	• C5-C6	• Axillary nerve
	• Upper trapezius (above horizontal)	• CN XI, C(2), 3, 4	• CN XI
Internal rotation	• **Subscapularis**	• C5-C7	• Subscapular nerve
	• **Teres major**	• C5-C7	
	• Anterior fibres of deltoid		• Lower subscapular nerve
	• Latissimus dorsi		
	• Pectoralis major		
	• Biceps		
External rotation	• **Infraspinatus**	• C5-C6	• Suprascapular nerve
	• Teres minor	• C5-C6	• Axillary nerve
Important point	• Winging of scapula may be because of serratus anterior or trapezius weaknesss. Winging during swan diving like posture (bending forward at waist and abduction of arms) is because of trapezius weakness, and winging during pushing of wall is because of serratus anterior weakness. • Every two degree motion at the glenohumeral joint, there is one degree of scapular rotation		
Elbow Flexion	• **Biceps brachii**	• C5-C6	• Musculocutaneous nerve
	• **Brachialis**	• C5-C6	• Musculocutaneous nerve
	• Brachioradialis	• C5-C6	• Radial nerve
Extension	• **Triceps brachii**	• C6-C8	• Radial nerve
	• Anconeus	• C7-C8	• Radial nerve
Supination	• **Supinator**	• C6-C7	• Posterior interosseous
	• Biceps brachii	• C5-C6	• Musculocutaneous nerve

Contd.

Table 21.1: Movements, their responsible muscles, and their innervations. Major innervations in bold, minor innervations in round brackets, main nerve from which branch has arrived in square brackets (Contd.)

Movements	Muscle	Segmental innervations	Peripheral nerve
Pronation	• **Pronator quadrates**	• C7-C8	• Anterior interosseous
	• **Pronator teres**	• C6-C7	• Median nerve
Wrist flexion	• Flexor carpi radialis	• C6-C7	• Median nerve
	• Flexor carpi ulnaris	• C7-T1	• Ulnar nerve
Extension	• Extensor carpi radialis longus	• C6-C7	• Radial nerve
	• Extensor carpi radialis brevis	• C7-C8	• Posterior interosseous
	• Extensor carpi ulnaris	• C7-C8	• Radial nerve
Adduction	• Flexor carpi ulnaris	• C7-T1	• Ulnar nerve
	• Extensor carpi ulnaris	• C7-C8	• Radial nerve
Abduction	• Flexor carpi radialis	• C6-C7	• Median nerve
	• Extensor carpi radialis longus	• C(5), C6, C7	• Radial nerve
	• and brevis	• C7-C8	• Radial nerve
Hands and fingers			
Flexion at PIP joint	• Flexor digitorum superficialis (FDS)	• C8-T1	• Median nerve
Flexion at DIP joint	• Flexor digitorum profundus (FDP)	• C8-T1	• Median nerve
Flexion at MCP joint	• **Dorsal and palmar interossei**	• C8-T1	• Deep palmar branch of ulnar nerve
	• Lumbricals	• C8-T1	• Deep palmar branch of ulnar nerve
Extension at MCP joint	• Extensor digitorum communis	• C7-C8	• Posterior interosseous nerve
	• Extensor indicis proprius	• C7-C8	• Posterior interosseous nerve

Contd.

Table 21.1: Movements, their responsible muscles, and their innervations. Major innervations in bold, minor innervations in round brackets, main nerve from which branch has arrived in square brackets (Contd.)

Movements	Muscle	Segmental innervations	Peripheral nerve
Extension at PIP joint	• Extensor digiti minimi	• C7-C8	• Posterior interosseous nerve
	• **Dorsal and palmar interossei**	• C8-T1	• Deep palmar branch of ulnar nerve
	• **Lumbricals**	• C8-T1	• Deep palmar branch of ulnar nerve
	• Extensor digitorum communis	• C7-C8	• Posterior interosseous nerve
Adduction of fingers	• Palmar interossei	• C8-T1	• Deep palmar branch of ulnar nerve
Abduction of fingers	• Dorsal interossei	• C8-T1	• Deep palmar branch of ulnar nerve
Abduction of little finger	• Abductor digiti minimi	• C8-T1	• Ulnar nerve
Abduction of thumb	• **Abductor pollicis longus**	• C7-C8	• Posterior interosseous nerve [radial nerve]
	• **Abductor pollicis brevis**	• C8-T1	• Median nerve
Adduction of thumb	• **Adductor pollicis**	• C8-T1	• Deep palmar br of ulnar nerve
Flexor of thumb	• **Flexor pollicis longus (distal phalanx)**	• C8-T1	• Anterior interosseous [median nerve]
Extensor of thumb	• **Extensor pollicis longus (distal phalanx)**	• C7-C8	• Posterior interosseous nerve [radial nerve]
	• Extensor pollis brevis (proximal phalanx)	• C7-C8	• Posterior interosseous nerve [radial nerve]
Opposition of thumb	• Opponens pollicis	• C8-T1	• Median nerve
Opposition of little finger	• Opponens digiti minimi	• C8-T1	• Ulnar nerve
Quiet inspiration	• **Diaphragm (vertical)**	• C3-C5	• Phrenic nerve

Contd.

Table 21.1: Movements, their responsible muscles, and their innervations. Major innervations in bold, minor innervations in round brackets, main nerve from which branch has arrived in square brackets (Contd.)

Movements	Muscle	Segmental innervations	Peripheral nerve
Deep insp- iration	• Intercostals (Transverse diameter) • Scalenii • Sternocleido- mastoid • Other muscles of shoulder, clavicles and scapula	• T1-T12	• Intercostal nerve [directly from nerve roots]
Important localizing point	• **Tidal percussion used to know the diaphragmatic excursion**		

Thoracic Spine

Extension	• Longissimus capitis • Iliocostalis tho- racis • Spinalis thoracis • Semispinalis thoracis • Rotatores thoracis	• All are sup- plied by dor- sal primary divisions of spinal nerves	
Lateral be- nding and rotation	All above except spinalis thoracis		

Lumbar spine

Extension	• Interspinales • Multifidus • Longissimus lumborum • Iliocostalis lumborum	• Dorsal pri- mary divisions of spinal nerves	
Lateral flexion	• Intertransver- sarii lateralis • Quadratus lum- borum	• Ventral pri- mary division of spinal nerves	

Contd.

Table 21.1: Movements, their responsible muscles, and their innervations. Major innervations in bold, minor innervations in round brackets, main nerve from which branch has arrived in square brackets (Contd.)

Movements	Muscle	Segmental innervations	Peripheral nerve
Rotation	• Intertrans-versarii mediales • Iliocostalis lum-borum • Multifidus	• T12, L1 • Dorsal primary divisions of spinal nerves • Dorsal primary divisions of spinal nerves • Dorsal primary divisions of spinal nerves	
Important point	Spinal muscles are examined enmasse.		

Abdomen

Lifting head and trunk in recumbent position	• Rectus abdominis • Transverses abdominis • Obliqui	• T5-T12 • T7-L1 • T7-L1	• Intercostal nerves • Intercostal, ilioinguinal, iliohypogastric • Intercostal, ilioinguinal, iliohypogastric
Important localizing point	**Beevor's sign: Umbilicus moves in downward direction if there is weakness of T9-T12 roots.**		
Pelvis	Watch for cremasteric, bulbocavernosus, and anal wink reflex		

Hip

Flexion	• **Iliopsoas** • Rectus femoris • Sartorius • Tensor fascia lata	• Iliacus from L2-L4, Psoas from L1-L4 • L2-L4 • L2-L3 • L4-S1	• Femoral nerve Psoas from lumbosacral plexus • Femoral nerve • Femoral nerve

Contd.

Table 21.1: Movements, their responsible muscles, and their innervations. Major innervations in bold, minor innervations in round brackets, main nerve from which branch has arrived in square brackets (Contd.)

Movements	Muscle	Segmental innervations	Peripheral nerve
Extension	• **Gluteus maximus** • Gluteus medius, minimus, hamstrings, hip adductors	• L5-S2	• Superior gluteal nerve • Inferior gluteal nerve
Adduction	• **Adductor longus, brevis, magnus** • Iliopsoas when hip is flexed • Gracilis	• L2-L4 • L2-L4	• Obturator nerve • Obturator nerve
Abduction	• **Gluteus medius, minimus** • **Tensor fascia latae** • Sartorius • Superior gluteal nerve	• L4-S1 • L4-S1 • L2-L3 • Superior gluteal nerve	• Femoral nerve
External rotation	• **Gluteus maximus** • Obturator internus • Obturator externus • Sartorius	• L5-S2 • L5-S1 • L3-L4 • L2-L3	• Inferior gluteal nerve • Nerve to obturator internus • Obturator n. • Femoral n.
Internal rotation	• **Gluteus medius, minimus** • **Tensor fascia latae**	• L4-S1 • L4-S1	• Superior gluteal nerve • Superior gluteal nerve
Knee flexion	• **Biceps femoris** • **Semitendinosus** • **Semimembranosus** • Popliteus • Gracilis • Sartorius • Gastrocnemius	• L5, S1-2 • L5, S1-2 • L5, S1-2 • L5-S1 • L2-L4 • L2-L3 • S1-S2	• Sciatic nerve • Sciatic nerve • Sciatic nerve • Tibial nerve • Obturator nerve • Femoral nerve • Tibial nerve

Contd.

Table 21.1: Movements, their responsible muscles, and their innervations. Major innervations in bold, minor innervations in round brackets, main nerve from which branch has arrived in square brackets (Contd.)

Movements	Muscle	Segmental innervations	Peripheral nerve
Extension	• **Quadriceps femoris** (rectus femoris, vastus lateralis, medialis, intermedialis)	• L2-L4	• Femoral nerve
Ankle			
Plantar-flexion	• **Gastrocnemius** • **Soleus** • Tibialis posterior • Peroneus longus • Peroneus brevis	• S1-S2 • S1-S2 • L5-S1	• Tibial nerve • Tibial nerve • Tibial nerve
Dorsi-flexion	• **Tibialis anterior** • Extensor digitorum longus • Extensor hallucis longus • Peroneus tertius	• L4-L5 • L5-S1 • L5	• Deep peroneal nerve • Deep peroneal nerve • Deep peroneal nerve
Inversion	• **Tibialis posterior** (chiefly in plantarflexed position) • Tibialis anterior (chiefly in dorsiflexed position)	• L5-S1 • L4-L5	• Tibial nerve • Deep peroneal nerve
Eversion	• **Peroneus longus** • **Peroneus brevis** • **Peroneus tertius** • Extensor digitorum longus	• L4-L5, S1 • L4-L5, S1 • L4-L5, S1	• Superficial peroneal • Superficial peroneal • Deep peroneal nerve
Important localizing point	• **Foot drop's important cause is L5 radiculopathy.**		
Toes			
Extension	• Extensor digitorum longus (EDL)	• L5-S1	• Deep peroneal nerve

Contd.

Table 21.1: Movements, their responsible muscles, and their innervations. Major innervations in bold, minor innervations in round brackets, main nerve from which branch has arrived in square brackets (Contd.)

Movements	Muscle	Segmental innervations	Peripheral nerve
	• Extensor digitorum brevis (EDB)	• L5-S1	• Deep peroneal nerve
	• Extensor hallucis longus (EHL)	• L5	• Deep peroneal nerve
	• Extensor hallucis brevis (EHB)	• L5-S1	• Deep peroneal nerve
Flexion	• Flexor digitorum longus and flexor hallucis longus	• L5-S1	• Tibial nerve
	• Flexor digitorum brevis and flexor hallucis brevis	• S1-S2	• Medial plantar nerve
Important localising point	• **EHL weakness is the hallmark of L5 weakness**		

- **Reflexes:**
  - Superficial (abdominal, cremastric). These are absent in corticospinal lesions.

Table 21.2: Important superficial reflexes

Reflex	Localising value
Cremasteric reflex	Afferent by femoral branch of genitofemoral nerve and ilioinguinal (L1-L2), efferent by genital branch of genitofemoral nerve (L1-L2)
Bulbocavernosus/ Bulbospongiosus reflex	Pudendal nerve (S2-S4)
Superficial anal reflex	Inferior hemorrhoidal nerve (S2-S5)
Plantar reflex (its pathological variant is Babinski reflex)	Afferent S1, Efferent L5, S1

  - Deep (muscle) tendon reflexes: DTRs may be hypoactive or hyperactive. These are absent in lower motor neuron, increased in UMN lesions and anxiety, and pendular in cerebellar lesions.

Table 21.3: Important deep tendon reflexes (hyperreflexia occurs if lesion is above these localising value)

Reflex	Localising value
Pectoral	C5-T1
Rhomboideus	C5
Biceps	C5,6
Triceps	C7
Supinator/brachioradialis	C5
Finger flexor reflex (Wartenberg's sign)	C8-T1
Patellar/knee reflex	L2-L4
Achilles reflex (ankle jerk)	S1

- **Pathological reflex:** For example, Babinski reflex, Hoffman's reflex. These appear in lesions of corticospinal tracts above their localizing value.

Table 21.4: Important pathological reflexes

Reflex	Localising value
Babinski	Afferent S1, Efferent L5, S1
Hoffman's (not always pathologic)	C5
Trommer	C5
Grasp reflex	Contralateral frontal lobe
Palmomental reflex	Contralateral frontal lobe

Table 21.5: Reflex patterns of different neurologic disorders[2]

Site or type of lesion	Muscle stretch reflex (DTRs)	Superficial reflex	Pathologic reflex	Associated movements
NM junction	Normal or decreased	Normal	Absent	Normal
Muscle	Usually normal	Normal	Absent	Normal
Peripheral nerve	Decreased to absent	Normal, or decreased to absent	Absent	Normal
Cortico-spinal tract (UMN lesion)	Hyperactive	Decreased/absent	Present	Pathologic associated movements present
Extrapy-ramidal system	Normal	Normal	Absent	Normal associated movements absent

Contd.

Table 21.5: Reflex patterns of different neurologic disorders[2] (Contd.)

Site or type of lesion	Muscle stretch reflex (DTRs)	Superficial reflex	Pathologic reflex	Associated movements
Cerebellum	Pendular	Normal	Absent	Normal
Psychogenic	Normal or increased	Normal or increased	Absent	Normal or bizarre
Periodic paralysis	Temporarily decreased	Normal	Absent	Normal

- **Abnormal movements:** Movement disorders may be hypokinetic or hyperkinetic. Hypokinetic disorders include Parkinson's disease, multisystem atrophy, progressive supranuclear palsy, corticobasal syndrome, diffuse Lewy body disease, Wilson's disease. Hyperkinetic disorders include tremors (essential tremor most common), chorea, athetosis, dystonia, hemiballismus, dyskinesias, orofacial dyskinesias, myoclonus, asterixis, myorhythmia, ticks, fasciculations, myokymia, spasms, etc.
- **Sensory examination:** It should be examined. Hemisensory loss may occur in parietal lesions, thalamic lesions, spinal lesions causing unilateral compression.
- **Coordination:** Disturbance of coordination in absence of weakness is seen in cerebellar.
- **Gait:** *See* chapter on gait.

Investigations

1. **CT brain:** This is ordered in acute development of hemiplegia or monoplegia, to look for intracranial haemorrhage, tumour bleed, chronic SDH, etc. Acute stroke may have normal CT picture; chronic one, however, may show hypodense infarct.
2. **MRI brain:** This is done when an intracranial cause is suspected. Space occupying lesions are well seen on it. DWI sequence is investigation of choice in acute **stroke.**
3. **Lumbar puncture:** It can be done in suspected cases of meningitis, encephalitis.
4. **Electrodiagnostic studies:** It includes nerve conduction studies and needle electromyography. It is done when

peripheral nerve disorders are suspected. It helps in delineating lesion site on the nerve, prognosis, and progression of the lesion. These are also helpful when examination is limited by pain and poor effort of the patient.

Important Points

Weakness may be classified as:

- Generalized or localized
- Symmetric/Asymmetric: Symmetric weakness occurs usually in extradural and myogenic causes, while asymmetric weakness usually occurs in intradural and neurogenic causes.
- Distal to proximal/proximal to distal: Former one occurs in neurogenic causes (UMN type), while the latter one occurs in myogenic (LMN type).
- UMN or LMN: Extradural lesions produce early UMN signs, with LMN signs at 1–2 involved regions. Intramedullary lesions produce late UMN signs and wide LMN signs because of involvement of anterior horn cells.
- Ascending or Descending. Ascending one occurs in extramedullary lesion, and descending one occurs in intramedullary lesion.
- Cruciate weakness occurs in foramen magnum lesions.
- Quadriparesis is mostly because of spinal cord disease, peripheral neuropathy, NM junction disorder, or a myopathy.
- In amyotrophic lateral sclerosis there is combined involvement of corticospinal tracts (because of cortex involvement) and anterior horn cells because of which patient will present with weakness, wasting, and hypertonia with increased reflexes in both upper limbs and lower limbs.
- UMN signs (extensor plantar) with hypotonia may occur in combined brainstem lesions and cerebellar lesions (Table 21.6).

Differential Diagnosis

Based on the site of level of lesion

1. **Cortical and subcortical lesions:** These produce UMN type of weakness. Cortical lesions include lesions involving

Table 21.6: Clinical difference between intramedullary, intradural—extramedullary and extradural pathologies

S. No.	Variables	Intramedullary	Intradural— extramedullary	Extradural
	Motor			
1.	UMN signs	Late	Early	Early
2.	LMN signs	Widespread due to AHC involvement	May be present at 1–2 involved segments	May be present at 1–2 involved segments
3.	Weakness	Descending	Ascending	Ascending
4.		Symmetric	Asymmetric	Symmetric/ asymmetric
	Sensory			
5.	Pain	Axial and funicular	Radicular	Localised or radicular
6.	Dissociated sensory loss	Present	Absent	Absent
7.	Sacral sparing	Present	Absent	Absent
8.	Joint/position sense involvement	Late	Early	Early
	Autonomic involvement			
9.	Bladder and bowel involvement	Early	Late	Late
	Local examination			
10.	Tenderness	Late/absent	Late/absent	Early, localised
11.	Gibbus	Absent	Absent	Present

motor cortex. Deficits will be according to their representation on the motor homunculus. Subcortical lesions involve internal capsule. Patients will have equal involvement of contralateral limbs. Both of these may have contralateral facial palsy.

Infratentorial lesions are because of brainstem involvement leading to lesions of pyramidal fibres. Pattern of deficits will depend on lesion supra or infra to the pyramidal decussation.

2. **Spinal lesion:** These produce LMN weakness at the level of lesion, and UMN weakness below it.

3. **Peripheral nerve lesion:** These produce LMN type of weakness. Weakness and wasting is equal in this.

4. **Neuromuscular junction disorder:** These produce diurnal variation of weakness. Presentation is of LMN type.

5. **Muscle disorder:** Presentation is of LMN type. Here, weakness is out of proportion to the atrophy.

6. **Disuse atrophy:** Here, atrophy is out of proportion to the weakness.

Based on the Etiology

1. **Vascular causes:** Stroke is the leading cause of vascular causes of motor deficit. Other causes may be intracranial hemorrhage, vasospasm following subarachnoid hemorrhage, endarteritis following infections.

2. **Space occupying lesions:** Tumours, abscesses, cysts may occur at any of the above mentioned sites.

3. **Infection:** Meningitis, encephalitis may produce motor deficits by causing endarteritis, leading to ischemia of the motor pathways. Anterior horn cell involvement, e.g. polio infection may also produce permanent motor deficit.

4. **Trauma:** Head, spine, neural, muscular injuries may lead to their motor deficits. These may present immediately, or delayed due to chronic SDH, hydrocephalous, abscesses, etc.

5. **Metabolic:** These include diseases like liver failure, kidney failure, glycogen storage diseases, etc. These may lead to acute motor deficits or muscular dystrophies.

6. **Autoimmune/inflammatory disorders:** These include acute disseminated encephalomyelitis, Guillain-Barré syndrome,[3] transverse myelitis, myasthenia gravis syndrome, Lambert-Eaton syndrome, etc.

7. **Others:** Todd's paralysis may occur post seizure. Ion abnormalities, idiopathic cellular dysfunction may cause periodic paralysis.

Table 21.7: Causes of compressive myelopathy		
Intramedullary	*Intradural-extramedullary*	*Extradural*
Tumours: Astrocytoma, ependymoma, dermoid, teratoma, hemangioblastoma	Tumours: Schwannoma, neurofibroma, arachnoid cyst, leukemic infiltration	Disc protrusion
Syringomyelia	Tubercular meningitis	OPLL, ossified ligamentum flavum, osteophytes
Hematomyelia	AVM Sarcoidosis	Trauma Developemental: AAD, kyphosis, scoliosis Infections: tuberculosis, abscess Tumours: Osteomas and other bony tumours, metastasis, hemangiomas Hematoma

REFERENCES

1. Medical Research Council. Aids to the examination of the peripheral nervous system, Memorandum no. 45. In: Office HMsS, ed. London 1981.

2. Campbell WW. Seventh ed. Philadelphia, USA: Wolters Kluwer/ Lippincott Williams & Wilkins; 2013.

3. Dimachkie MM, Barohn RJ. Guillain-Barré syndrome and variants. *Neurologic clinics.* 2013;31(2):491.

22 | Movement Disorders

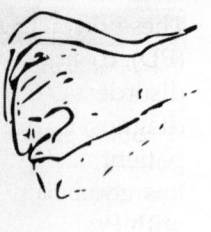

```
                    Movement disorders
         │                              │
    History                        Examination

  Onset and duration               General
  progression
                                   Higher mental
  Age                              function

  Anatomical part                  Cranial nerves

  Hyperkinetic/                    Sensorimotor
  hypokinetic
                                   Cerebellar
  Intention/resting
                                   Gait
  Rhythmic/non-rhythmic
                                   Abnormal
  Stereotyped/astereotyped         movement exam

  Pre-movement urge

  Gait

  Shock-like movements

  Provoking and relieving factors

  Persistence in sleep

  Motor impersistence

  Non-motor symptoms

  Trauma

  Metabolic abnormalities

  Drugs

  Cognitive and
  psychiatric abnormalities

  Family history
```

These disorders include tremors, chorea, Parkinson's disease (PD), dystonia, and their syndromes. Diagnosis of movement disorders is purely clinical. It is important to establish a correct diagnosis in order to impart best possible treatment to the patient and prognosticate him, for instance, a patient of PD has good response with intervention compared to a patient with Parkinson plus syndrome.

Terminology

1. **Tremor:** Involuntary **rhythmic** sinusoidal movements of one or more body parts.
2. **Tics: Stereotyped** recurrent movements. The movements can be controlled voluntarily for short time. There is a sense of discomfort or urge relieved by the movement, e.g. itch or scratch.
3. **Chorea:** Abrupt, unpredictable, and non-rhythmic movements. This changes from one body part to another, and there occurs **randomly flowing jerks.**
4. **Ballism:** This can be considered under the rubric of chorea. There is uncontrollable, large amplitude, unpredictable, proximal muscle movements.
5. **Myoclonus:** Sudden brief **shock-like** involuntary movements can be positive (due to muscular contractions), or negative (due to loss of muscular tone).
6. **Dystonia:** Abnormal co-contraction of antagonistic muscles, leading to abnormal postures.
7. **Athetosis:** Distal mobile dystonia, manifested as slow writing movements of fingers.
8. **Ataxia:** It refers to the clumsy and poorly organised movements, and may affect speech, dexterity or gait.

History

- **Onset, duration, and progression:** Mostly movement disorders are degenerative and chronic. Atypical PD has onset of about one year only. Acute onset movement disorders can be due to drugs, infections, trauma, ischemia.
- **Age:** Chorea and physiological tremor may start in childhood, dystonia may start in teens, PD starts in middle age, dystonic tremor and task-specific tremor starts in adulthood. Noting age is also important for therapeutic

purposes as advanced age > 80 years is generally considered a contraindication for neuromodulation surgery.

- **Anatomical body part:** Patient should be enquired which body part he is having abnormal movements. This may be from head to toe. Focal disorders affect one body part, regional affect two contiguous body parts, and generalised ones include axis parts, or both sides of body, or both. Proximal or distal involvement should also be enquired.

- **Hyperkinetic/hypokinetic:** Patients will complain of either decreased or increased movements. One should also enquire "whether they are absent". Table 22.1 shows the types of these movements. This classification helps the clinician a lot in progressing towards the diagnosis.

- **Intention/resting:** Intentional tremors occur in cerebellar disorders, dystonia, and essential tremors; while resting tremors are seen in PD. Myoclonus is generally seen in rest state, but can be seen in working state or get worsened by it.

- **Rhythmic/non-rhythmic:** Rhythmic movements are tremors. Rest all movements are non-rhythmic. When myoclonus occurs in series, it may look like rhythmic.

- **Sterotyped/astereotyped:** Sterotyped (always same) movements are tics. Rest are asterotyped.

- **Pre-movement urge:** This condition occurs in tics. Patient may complain of a sense of discomfort/urge before the movement, and this vanishes only after doing it. This is characteristic of tics.

- **Gait:** Patient should be enquired about the slowness, or instability during the gait. Slowness is seen in PD, while instability is seen in ataxias.

- **Shock-like movements:** These are characteristic of myoclonus. These can be positive (due to increased muscular tone), or negative movements (due to loss of muscular tone), e.g. bouncy gait (https://www.youtube.com/watch?v=fhv3 fhut26Q).

- **Provoking factors:** Myoclonus can be provoked by tactile stimuli, known as reflex myoclonus. Musician's dystonia, writer's cramp are triggered with their actions. Orthostatic tremor is triggered by standing.

- **Relieving factors:** Patient may tell, they can apply some sensory stimulus to relieve the movement, for example, in dystonia.

- **Persistence in sleep:** Sleep causes improvement in dystonia, chorea, and tremors, except the palatal tremor.

- **Motor impersistence:** There is waxing and waning of power in chorea, e.g. milkmaid's grip (https://www.youtube.com /watch?v=R2zCH18e0n0).

- **Non-motor symptoms:** Drooling of saliva may be seen in PD; orthostasis may be seen in multiple system atrophy; depression may be seen in any of the disorders, more common in PD; psychiatric symptoms may be seen in Huntington's disease. Speech may be abnormal in ataxias.

- **Trauma:** This should be enquired as post-traumatic tremors are well known.

- **Metabolic abnormalities:** Liver, kidney disease, Wilson's disease should be enquired about. These can present with toxin accumulation, and secondary forms of movement disorders.

- **Drugs:** Provoking drugs and relieving drugs should be asked. Drugs associated with tremors include alcohol, amiodarone, phenytoin, lithium, sympathomimetics, etc. Patients with levodopa responsiveness with drug induced dyskinesias are the best candidates for deep brain stimulation/lesioning. On and off period of levodopa should also be asked. The initial on period with levodopa in PD is 3–4 hours, which decreases to about 30 min in advanced stages.

- **Recent infections:** Sydenham's chorea develops 4–8 weeks to 6 months after a Group A β-hemolytic streptococcal infection.

- **Cognitive and psychiatric abnormalities:** These may be associated with movement disorders, or with their drugs (psychosis and impulse control disorders with dopaminergic agonists).

- **Family history:** Essential tremor runs in families, and incidence of affection in first degree relatives ranges from 17–96%.[1,2] Huntington's chorea is autosomal dominant.

Table 22.1: Classification of movement disorders[3]

Movements		Terminology	Symptoms	Disease
Decreased		Bradykinesia Hypokinesia	Slowness Decreased amplitude	Parkinsonism Cerebellar disease, pyramidal disorders
Increased	Jerky	Myoclonus	Sudden shock like movements (focal/ multifocal/ segmental/ generalised)	Focal—cortical; Generalised— subcortical
		Chorea	Abrupt, unpredictable, and nonthymic Waxing and waning of power	Sydenham's chorea
		Ballism	Large amplitude, uncontrollable, proximal muscle movements.	Hemiballism: Contralateral subthalamic nucleus Bilateral: Metabolic abnormalities
		Tics	Stereotyped recurring movements, with rising discomfort prior to the movement	Tourette's syndrome, autism, mental retardation, Rett syndrome, psychosis, congenital blindness, and deafness.
		Ataxia	Clumsy and poorly organised movements	Spinocerebellar ataxia, ataxia telengiectasia Friedrich's ataxia, etc.
	Non-jerky	Dystonia	Abnormal postures	Primary (DYT mutations), secondary (drugs, infections, ischemia, trauma, etc.)
		Tremor	Involuntary, rhythmic, sinusoidal movements	PD, essential tremors, drug induced, etc. (explained in Table 22.2)

Examination

- General examination: Examine spontaneous blinking of eyes, facial expressions. These automatic movements are decreased (hypomimia) in PD.
- Higher mental functions: Perform higher mental functions to rule out mood disorders, and associated syndromes.
- Cranial nerves: Examine all cranial nerves. These are generally normal in movement disorders.
- Sensorimotor examination:

  Examine spasticity and rigidity: Spasticity is velocity dependent, and is seen when done quickly. In it, there is resistance only at the beginning, like a *clasp knife*. It is seen in pyramidal disorders. In rigidity, there is uniform resistance throughout the range of movement, like a *lead pipe*, and is seen in Parkinson's disease. *Cogwheeling* is another type of rigidity, in which there is hand-break like dystonic tremor throughout the range of movement. This occurs in essential, dystonic tremor, and in PD.
- Cerebellar examination: Perform repetitive alternating movements like rapid supination and pronation of hand. Disorderly performance of it is known as dysdiadecho-kinesia, seen in cerebellar hemispheric disorders. Slowness of movements (bradykinesia) is seen in cerebellar and pyramidal disorders. Fatiguing and decrement of it (hypokinesia) is seen in parkinsonism.
- Gait examination: Ask the patient to walk in front of you. A Parkinsonian gait is of short shuffling steps, with difficulty to control when asked to stop. Reduced arm swing (another example, hypomimia) is another feature seen in PD. Bouncy gait occurs in negative features of post-anoxic myoclonus. Gait is also abnormal in ataxias. See the chapter on gait for detailed examination.
- Abnormal movement examination: Examine the abnormal movement which is occurring to the patient.
  - Tremors: Examine whether these are:
    - Resting/intention: Resting occurs in essential tremors, PD. Intention tremor occurs in cerebellar ones, dystonic tremors.
    - Symmetrical/asymmetrical: Physiological tremor and ET are symmetrical. Rest are asymmetrical.

> ➤ Low or high frequency: Tremors of hands less than 6 Hz and more than 11 Hz are always pathological. Rubral tremors are of low frequency—2–5 Hz. Physiological tremors are of high frequency—8–12 Hz. PD tremors are 4–8 Hz.
> ➤ What is moment of occurrence: Table 22.2 classifies tremors according to their moment of occurrence.
> ➤ Effect of loading: Pathological Essential tremors abolish on loading of the limb, while physiological tremors do not get abolished.

– Tics: Examine the stereotype nature of the movement. Patient can be told to mimic the movement. Sometimes, it is absent in the examination room, for which videotaping can be done. It is generally present in facial region, neck, or upper arm.

– Chorea: Examine whether it is focal or generalised. It may be athetosis with slow writing movements of hands or feet, or wide flinging movements of arms like in hemiballismus.

– Myoclonus: Examine whether the myoclonus is focal, multifocal, segmental, or generalised.

– Dystonia: Examine whether the dystonia is focal, segmental, multifocal, or generalised. There are several scales available to measure dystonia, for example, Burke-Fahn-Marsden dystonia rating scale (BFMDRS). These are required to quantitatively measure the disease burden percentage to see the response of therapy, and when surgery is considered.

– Postural instability: **Retropulsion pull test:** It is done to determine postural stability in PD. Patient is asked to stand up with eyes opened, with feet-shoulder width apart. The examiner stands behind the patient. After explaining him, tell that you are going to pull him and will support him if he falls. Then, give a sufficient pull to disbalance him, and observe whether he can balance himself. Scoring is given in the following form, according to UPDRS score

> ➤ 0 = recovers independently may take 1 or 2 steps or an ankle reaction;
> ➤ 1 = three steps or more backward but recovers independently;

> 2 = retropulsion, needs to be assisted to prevent fall;
> 3 = very unstable, tends to lose balance spontaneously;
> 4 = unable to stand without assistance
- Auscultation: At times, one can hear thumping sounds in orthostatic tremor when auscultating over legs.[4]

Investigations

1. **MRI brain:** It should always be done to rule out structural lesions. One can see focal lesions like neurodegeneration with brain iron accumulation (NBIA). Huntington's disease may have caudate atrophy and boxing of lateral ventricles, and frontal lobe atrophy. Sydenham's chorea may show transient swelling of globus pallidus and stritum. Figure 22.1 is showing MRI T2 image with NBIA in bilateral basal ganglia, in a patient with generalised dystonia.

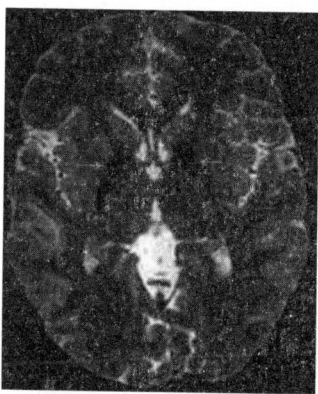

Fig. 22.1: Bilateral basal ganglia NBIA

2. **TRODAT scan:** Technetium 99m tropane derivative-1 analysis can be used to image the dopaminergic system, and to identify the PD in early stages. It is not taken up by thyroid gland and can be analysed within hours.
3. **DYT1 gene mutation:** It is present in primary dystonia. It has prognostic value also, as the response of deep brain stimulation, and lesioning is better when this is positive.
4. **Serum copper, total urine copper, and ceruloplasmin levels:** These are needed to rule out Wilson's disease, which can present with secondary form of dystonia.

5. **Huntington gene:** CAG repeat on chromosome 4 is the characteristic finding in HD.
6. **ASO titre:** These are elevated in Sydenham's chorea.
7. **SCA, ATM gene, frataxin gene mutations:** In clinically appropriate scenarios of ataxias.

Differential Diagnosis

1. **Tremors:** The differentials can be made on the basis of moment of occurrence, given by consensus statement of the Movement Disorder Society on Tremor, Ad Hoc Scientific Committee.[5] Table 22.2 is taken from the article by Abdo et al.[3]

Table 22.2: Classification of tremors according to moment of occurrence		
Moment of occurrence	*Features*	*Underlying disorder*
A. At rest	Best judged in a body part that is fully supported against gravity	Parkinson disease
B. With action postural	Occurs in body part that assumes a posture against gravity	Physiological; enhanced physiological (stress, endocrine disorders or intoxications); essential tremor
Kinetic		
a. Simple	Occurs during entire movement trajectory	Essential tremor
b. Intention	Progressively increases towards intended target	Cerebellar ataxia
c. Task-specific	Occurs only during specific activities	Dystonic writing tremor
Isometric	Occurs during voluntary muscle contractions against a stationary resistance	Physiological; associated with other types of tremor
C. Combinations	Various	Severe essential tremor; atypical parkinsonism; dystonic tremor; rubral (Holmes) tremor

Specific Tremor Disorders

- **Physiological tremors:** It starts in childhood, has a frequency of 8–12 Hz with low frequency proximally, and more distally, and is symmetrical.
- **Essential tremor:** It is the most common type of tremor. Starts generally in adulthood, worsens with age, has symmetrical action and postural tremors with a frequency of 6–8 Hz. It involves the upper limbs in about 95% cases, head in about 34%, and lower limbs in about 20%.[6] Etiology is not clearly known, but may be due to an underlying cerebellar or brainstem pathology.
- **Parkinsonian tremor:** Please see the following paragraph on PD.
- **Cerebellar tremor:** It is characterized by asymmetrical intentional tremor of low frequency (2–4 Hz), high amplitude tremors, accompanied by cerebellar signs.
- **Rubral tremor:** These are the slowest tremors with frequency of <4 Hz. These are resting tremors. They arise from damage to cerebellar and brainstem pathways.
- **Post-traumatic tremors:** These occur post head injury and is seldom alone. The manifestation depend upon the region involved, and thus can be resting, postural or intention.
- **Drug-induced tremors:** These are symmetrical, with low frequency, and has temporal relation with drug ingestion. Agents include alcohol, amiodarone, phenytoin, lithium, sympathomimetics, etc.
- **Dystonic tremor:** It is a jerky postural and action tremor present in the dystonic body parts. It is abolished on rest. The tremor is often reduced by sensory tricks.
- **Psychogenic tremor:** It can vary in frequency, amplitude depending upon the state of the patient. It often gets abolished with distraction. Even the tremor gets shifted from one body part to another.
- **Chorea:** It is manifested by random involuntary fidgety movements that move from one body part to another (https://www.youtube.com/watch?v=HOalYWvVLU8). It may be focal or generalised. They can be hereditary like Huntington's disease (begins in 3rd to 5th decade of life), Sydenham's chorea, Friedriech's ataxia, etc. or due to metabolic reasons like Wilson's disease, liver or kidney disease, or due to drugs.

- **Huntington's disease:** It is the commonest form of hereditary chorea, manifesting as autosomal dominant. It presents in the 3rd to 5th decade of life, and is characterised by symmetrical chorea, gait unsteadiness (ataxia is unusual), cognition abnormalities, psychiatric changes, and loss of bulbar control. Motor impersistence is a classical feature which can be tested by asking the patient, and checking his inability to protrude his tongue, or by checking the milkmaid type grip when asked to hold your finger. MRI may show caudate atrophy and boxing of lateral ventricles, and frontal lobe atrophy. CAG repeat on chromosome 4 is the usual defect.
- **Sydenham's chorea:** It is the most common form of childhood chorea; develops by 8–9 years of age; after 4–8 weeks, sometimes up to 6 months after a Group A β-hemolytic streptococcal infection. Clinical features are chorea, tics, psychiatric disturbances like OCD. MRI may show transient swelling of globus pallidus and stritum. Elevated ASO titres help in distinguishing from other types of chorea.
- **Tardive dyskinesia/chorea:** It occurs in the setting of typical neuroleptics like haloperidol, though it may occur with atypical ones also. Other drugs include metoclopromide, promethazine.
- **Myoclonus:** It is a sudden arrhythmic startle response, which may be positive (hypnic jerks), or negative (asterixis). It may be epileptic or non-epileptic. It can occur in epilepsy, Creutzfeldt-Jakob disease, cortical, basal ganglionic, or cerebellar degeneration, or PD.
- **Dyskinesia:** It is abnormal arrhythmic movements, look like chorea or dystonia. It involves limbs, or trunk, or both. It can be abdominal (belly's dance), levodopa induced, tardive (after neuroleptics), or paroxysmal.
- **Tics:** Primary tics start in childhood, and adult onset are due to secondary causes. Patient has a premonitory urge which gets relieved only by during the movement, which can be in motor, or vocal forms. These movements can be temporarily suppressible, but rebound with increased severity. Primary tics disorders include Tourette's syndrome, secondary causes of tics include infections, neuroleptic drugs exposure (antiepileptic drugs, neuroleptics), toxins, or trauma.

- Tourette's syndrome: It develops before 18 years of age. By definition, tics must be present for a year, and slowly builds up. It is commonly associated with OCD and ADHD. It is transmitted by autosomal dominant mode.

Parkinson's disease: Tremor, rigidity, bradykinesia, and postural instability are the cardinal features of PD. Parkinsonism include bradykinesia with any of the other cardinal features. It starts in the fifth decade of life (if it starts before 50 years, then it is called young onset PD, while if it starts before 20 years, then it is called juvenile PD). It manifests with tremor, rigidity, and bradykinesia. Tremor is pin rolling type resting one with a frequency of 4–8 Hz. The tremor is asymmetrical, with the classical progressive involvement of one upper limb, ipsilateral lower limb, and then contralateral upper limb. The bradykinesia is the most reliable sign of disease severity. Patient manifests with shuffling gait, stooped posture, enbloc turning. He will also have difficulty in maintaining posture, during the pull test. Rigidity is lead pipe type. Causes include idiopathic (most common), environmental exposure (pesticides, drinking well water), and genetic ones (PARK1, PINK1, LRRK2 mutations). Presence of autonomic dysfunction, cerebellar deficits, corticospinal signs, dementia are signs of Parkinson's plus syndromes.

Parkinson's plus syndromes include dementia with Lewy bodies (progressive parkinsonism, dementia, and hallucinosis), Progressive supranuclear palsy (progressive parkinsonism, gait instability and gaze palsies), corticobasal degeneration (parkinsonism with cortical features including apraxia, and alien hand phenomenon), multiple system atrophy (parkinsonism with dysautonomia—impotence, diaphoresis, orthostatic hypotension, and incontinence).

Akathisia: It is characterised by inner feelings of restlessness, like inability to keep legs still. The movement relieves the restlessness. Restless leg syndrome (RLS) is an akathitic disorder. Causes may include drugs—dopamine blockers, SSRIs, calcium channel blockers, tricyclic antidepressants, etc.

Stereotypies: Repetitive movements that mimic purposeful action, but performed outside the situation. It may be voluntary, or involuntary. Examples include jumping, facial grimacing, body rocking, etc. It may be seen in autism, Rett's syndrome, schizophrenia, mental retardation, neuroleptic medications, etc.

Dystonia: It is the abnormal co-contraction of agonist and antagonistic muscles that result in abnormal postures/and or movements. It can be classified as in Table 22.3. It can also be classified as genetic or acquired. The most common heritable cause is DYT-1 mutation,[7] and this is also the most common cause of young onset generalised dystonia. Other heritable causes include DYT-2 to 15 mutations. Acquired causes include ischemia, hypoxia, trauma, infections (Creutzfeldt-Jakob disease, HIV, viral encephalitis), toxins (CO poisoning, cyanide, disulfram), metabolic diseases (hypoparathyroidism), ischemia (stroke), trauma, syringomyelia, peripheral nerve injury, Wilson's disease, drugs (anticonvulsants, fenfluramine) (Table 22.3).

Table 22.3: Clinical classification of dystonia			
According to	*Type*	*Involvement*	*Example*
Location	Focal	One region	Writer's cramp, cervical torticollis
	Segmental	Two contiguous body parts	Craniocervical dystonia
	Multifocal	More than two body parts in one side of body	Hemidystonia
	Generalised		
Age	Early onset	Less than 26 years	Heritable forms (DYT)
	Late onset	≥ 26 years	Acquired ones
Temporal course	Static Permanent		
Coexistence of other disorders			With myoclonus, parkinsonism, cognitive, psychiatric deficits

Wilson's disease: All patients with young onset movement disorders should be checked for Wilson's disease. There is involvement of liver and brain, with neurologic manifestations of parkinsonian symptoms, dystonia, ataxia, and dysarthria. It occurs secondary to ATPase gene mutation on chromosome 13.

Ataxia: There is defect in coordination, balance and speech. It may be due to damage to cerebellum, spinal cord, or peripheral nerves. It may be acquired, hereditary, or idiopathic-late onset cerebellar ataxia (ILOCA).

- **Acquired ataxia:** It may be due to head injury, stroke, meningitis/encephalitis, multiple sclerosis, nutritional deficiencies—vitamin B_{12}, chronic alcohol consumption, tumours, benzodiazepines.

- **Hereditory:** These are genetic syndromes and can be autosomal dominant, autosomal recessive, and X-linked ataxias.

  Autosomal dominant ataxias include **spinocerebellar ataxia** (SCA 1, 2, 3, 6, 7, 8). These have pyramidal signs like hyper-reflexia and spasticity. SCA-3 which is the most common has hyper-reflexia, spasticity, nystagmus, and dysarthria. It may also have levodopa responsive tremor. MRI may show pontocerebellar and spinal atrophy. **Episodic ataxias** occurs for brief duration. There are seven types, all of which are due to ion channel defect. EA-1 (potassium channel defect) lasts for seconds to minutes, and EA-2 (calcium channel defect) lasts up to a day. Patients may have vertigo, myokymia, migraines, blurred vision, tinnitus, and dysarthria.

  Autosomal recessive ataxias include the **Friedrich's ataxia (FA)**. It is childhood onset manifested by ataxia, weakness, and dysarthria. It can also have associated cardiomyopathy, and spinal abnormalities like scoliosis. **Ataxia telengiectasia (AT)** starts in childhood, and manifests with ataxia (telengiectasias develop later). As the age advances, patient may have hypotonia, areflexia, bradykinesia, and proprioceptive deficits. There is predisposition for development of neoplasms like acute lymphocytic leukemia, or lymphoma. AT has deficit in DNA repair pathway.

  Ataxia with isolated vitamin E deficiency: It mimics FA, and has childhood onset. It manifests with ataxia, areflexia, dysarthria, and bony abnormalities. It is caused by mutation in alpha-tocopherol transfer protein. Symptoms can be reversed and halted with vitamin E replacement.

  X-linked ataxias include fragile-X associated tremor ataxia syndrome (FXTAS) characterized by ataxia, intention tremor, bradykinesia, rigidity, polyneuropathy, and autonomic manifestations. MRI-T2 imaging may reveal bilateral symmetrical hyperintensities in middle cerebellar peduncles.

- **Idiopathic late onset cerebellar ataxia:** The reasons are unclear.

REFERENCES

1. Deng H, Le W, Jankovic J. Genetics of essential tremor. *Brain.* Jun 2007;130(Pt 6):1456–1464.
2. Busenbark K, Barnes P, Lyons K, Ince D, Villagra F, Koller WC. Accuracy of reported family histories of essential tremor. *Neurology.* Jul 1996;47(1):264–265.
3. Abdo WF, Van De Warrenburg BP, Burn DJ, Quinn NP, Bloem BR. The clinical approach to movement disorders. *Nature Reviews Neurology.* 2010;6(1):29–37.
4. Brown P. New clinical sign for orthostatic tremor. *Lancet.* Jul 29 1995;346(8970):306–307.
5. Deuschl G, Bain P, Brin M. Consensus statement of the Movement Disorder Society on Tremor. Ad Hoc Scientific Committee. *Mov Disord.* 1998;13 Suppl 3:2–23.
6. Elble RJ. Diagnostic criteria for essential tremor and differential diagnosis. *Neurology.* 2000;54(11 Suppl 4):S2–6.
7. Bressman SB. Dystonia: Phenotypes and genotypes. *Rev Neurol (Paris).* Oct 2003;159(10 Pt 1):849–856.

Gait ataxia

History

- Onset, duration and progression
- Age
- Stance
- Character
- Diurnal variation
- High steppage
- Foot drop
- Walking on irregular surface
- Cotton wool like sensations
- Maneuvering through narrow lanes
- Negotiating feet in chappals/sandals
- Slippage of chappals
- Threading needles
- Buttoning/unbuttoning
- Opening door knobs
- Writing
- Nystagmus
- Speech
- Vestibular symptoms
- Tremors
- Headache
- Rigidity and spasticity
- Fever, rash
- Drugs
- Trauma
- Family history

Examination

- MMSE
- Examination of cranial nerves
- Sensorimotor examination
- Cerebellar function examination
- Examination of gait

Gait: Gait problem is a very common complaint, and its analysis is a very useful tool for the neurologists and neurosurgeons to reach the diagnosis. Though both neurologic and orthopaedic problems may impair the gait of a person, we will focus only on neurologic entities.

History

- **Onset, duration and progression**: Acute onset ataxia may be due to intracranial infections, post-immunizations, haemorrhage inside the tumour, CVAs, injuries. Chronic ones include tumours, senile causes.

- **Age:** Congenital and hereditary ataxia (e.g. Friedreich's ataxia) are seen in childhood. Old age disorders are cautious senile gait, NPH, and parkinsonism.

- **Stance:** Ask whether he can maintain a stance both in daylight and night. If he sways in daylight with eyes open, then it means cerebellar cause is there. If he sways in night only and not in day, then it means sensory cause is there.

- **Character:** Ask whether he walks slowly, broad based, has tendency to sway, walks like a drunkard, sways side to side, increased swaying in dark/eyes closed, freezing, shuffling, or apraxia.

- **Diurnal variation:** Ask whether ataxia exacerbates in night time, or there is difficulty in washing face at night. This occurs due to impaired proprioceptive and joint-position sense in sensory ataxia, which worsens due to loss of visual input in the night.

- **High steppage:** Ask whether there is high steepage gait, and if so, then which first: Foot first (motor involvement), heel first (sensory involvement).

- **Foot drop:** Ask whether there is foot drop. In unilateral foot drop—L5 nerve root, peroneal nerve involvement. In bilateral: Charcot-Marie-Tooth disease, severe peripheral neuropathies, muscular dystrophies.

- **Walking on irregular surface:** Patient may tell that he has more difficulty in walking on irregular surface like cobblestones. This occurs in sensory ataxia.

- **Cotton wool like sensations** while walking: This occurs because of sensory problems.

- **Maneuvering through narrow lanes:** Ask the patient whether there is exaggerated difficulty in walking through narrow lanes, like boundaries of farms, or there is difficulty in walking at crowded places. This occurs due to impaired coordination because of cerebellar cause.

- **Negotiating feet in chappals/sandals:** This occurs due to impaired coordination, that may occur either due to cerebellar, or with motor involvement.

- **Slippage of chappals:** Ask if slippage of chappals/sandals occur. If it occurs without knowledge, the cause is sensory; while if it occurs with knowledge, the cause is motor involvement.

- **Threading needles:** This question we can ask from house-wives. In absence of motor weakness, difficulty in threading needles occurs due to impaired coordination due to cerebellar causes.

- **Buttoning/unbuttoning:** In absence of weakness, this complaint suggests cerebellar problem.

- **Opening door knobs:** Difficulty in performing rapid alternating movements like opening of door knobs points towards cerebellar ataxia.

- **Writing change:** Ask the patient whether he has experienced writing change. Macrographia suggests cerebellar cause, and micrographia suggests Parkinson's disease.

- **Nystagmus** (*see* Nystagmus chapter)

- **Scanning speech:** This occurs due to cerebellar causes.

- **Difficulty in feeding himself/drinking water:** Ask whether patient has difficulty feeding himself, and spills food or water while taking to the mouth: This may suggest incoordination due to cerebellar causes.

- **Vestibular symptoms:** Nausea, vomiting, vertigo, light headedness suggest vestibular causes.

- **Tremors:** Ask whether there is any tremor, and if so, then it is intentional (i.e. during a task) or at rest. Tremors due to cerebellar lesions are intentional, while due to PD are at rest.

- **Headache:** Ask the patient about signs of raised ICP like early morning headaches, blurred vision, projectile vomiting. This point towards an intracranial space occupying lesion.

- **Rigidity and spasticity:** Rigidity may occur due to Parkinson's disease. Ask the patient whether he feels any tightness or early fatigue for spasticity, that may occur due to motor weakness of upper motor neuron type.
- **Fever, rash:** These may occur in viral illness.
- **Drugs:** Enquire about advertent or inadvertent drug ingestion which may produce ataxia. Children may accidentally take those drugs, alcohol or household chemicals.
- **Recent trauma:** Trauma may cause cereberal contusions, vertebral artery dissection.
- **Family history:** It needs to be taken to link causes.

Examination

MMSE: To rule out dementia. If abnormal, then do higher mental function examination (*see* Headache chapter).

Examination of cranial nerves: Cranial nerve examination can help in localization of the lesion (*see* respective chapters). Papilledema points towards raised ICP. Vth, VIIth, VIIIth, IXth, Xth, XIIth nerves involvement point towards CP angle region.

Sensorimotor examination: Sensory examination abnormality can help in identifying sensory ataxia (*see* Sensory Impairment chapter). Similarly, motor examination can help in localisation of the lesion (*see* Weakness chapter)

Cerebellar function examination:[1] The abnormalities occur ipsilateral to the cerebellar lesion.

- **Dyssynergia:** This means absence of synergism between different components of a task. Ask the patient to perform finger nose test.

  Be seated in front of the patient. Keep your finger in front of him at an arm's length. Show him how to touch his finger to your fingertip and then to his nose (the forearm should be parallel to the ground). Ask him to repeat the task first from normal, then from abnormal side. In dyssynergia the act will be a jerky, disorganised and in erratic manner.

  Other test for examining this is heel-shin test where the patient has to slide his heel of one limb over the shin of other leg.

  Dyssynergia occurs due to cerebellar hemispheric lesions.

- **Dysmetria:** This means errors in judging distance, speed, power, and direction. The patient's finger, in the finger-nose test, may overshoot the target (hypermetria) or undershoot the target (hypometria). The finger may also go in wrong direction.

Dysmetria occurs due to cerebellar hemispheric lesions.

- **Agonist-Antagonist coordination (dysdiadochokinesia):** Show the patient to perform rapid alternating movements like supination and pronation of one hand over the other, then ask him to do (first normal and then abnormal side). Loss of this ability is suggestive of agonist-antagonist discordance.

  This also occurs with cerebellar hemispheric lesions.

- **Tremors:** Ask the patient to reach to a particular object. Irregular jerky movements perpendicular to the path of movement at the start, during the act, or at termination occurs due to cerebellar hemispheric cause.

  Titubation (jerking of the head and body) occurs in midline vermian lesions.

- **Hypotonia:** This occurs due to loss of cerebellar facilitation to the motor cortex. Stretch reflexes may also diminish or become pendular—there is to and fro movement of leg on tapping the quadriceps tendon.

- **Holmes Rebound phenomenon:** Sit in front of patient. Hold the patient's wrist, and ask him to flex the elbow. On suddenly releasing, the patient's forearm should flex a little and then extend a little again. Exaggeration of this phenomenon is seen in spasticity, while absence of this is seen in cerebellar hemispheric lesions (due to loss of antagonist muscle to contract)[2].

- **Dysarthria:** Vermian lesions produce the scanning speech.

- **Nystagmus:** *See* Nystagmus chapter

- **Examination of gait**
  - Stance: Width of base (first compensatory effort by patient is increase in the width)
  - In normal gait two phases: Stance phase (60%) and swing phase (40%)
  - Stance phase is divided into: Heel strike, mid stance, push off
  - Swing phase is divided into: Acceleration or initial swing, mid swing, deceleration

- Other components which needs to be seen: Pelvic tilt, pelvic rotation, lateral shift, width of base, stride length, Step length, arm swing, tremors or chorea.
- Examine width of base by standing behind the patient while making him walk. It is the horizontal distance between the feet at the time of double stance or double support.
- Stride length is the length between point of heel strike of one foot, to the point of the heel strike of the same foot.
- Step length is the length between heel strike of one foot to the heel strike of the other foot.
- Normal pelvic tilt is 5 degrees
- Normal pelvic rotation is approx 4 degrees anteriorly on the swing side, and approx 4 degrees posteriorly on the stance side
- Lateral shift is approx 1 inch towards the stance phase leg
- Pelvic tilt, pelvic rotation, lateral shift, in combination with knee flexion and knee, ankle and foot motion combines to maintain centre of gravity (COG) in both vertical and horizontal axes.
- In the anatomical position the COG lies anterior to S2 vertebra.

- Normal base width is 2–4 inches.
- Normally stride length of both legs are equal, and the step length of the both the left and right leg will be equal.
- Abnormality of any component of stance phase or swing phase or of other components suggest a pathology which should be identified by the examiner.
- Make the patient walk tandemly or hopping to make out subtle abnormalities.

Investigations

1. **CT/MRI head:** These are indicated when an intracranial cause is suspected.
2. **CSF examination:** It is indicated for finding or ruling out the infectious causes.
3. **Drug levels:** Specific drug levels may be ordered as suggested by history.

Specific Gaits

- **Cerebellar ataxic gait:** Broad based, lurching, titubating, may sway to either side, back or forward. To elicit abnormality in mild ataxia—make the patient to walk tandem. Other test to identify the subtle cerebellar ataxia is asking the patient to turn suddenly while walking. Present with eyes open and closed. With hemispheric lesion patient may sway towards the involved side. Star-shaped gait (Unterberger-Fukuda stepping test) may occur (it is also seen in vestibular lesion).

- **Sensory ataxia:** High steppage gait. Heel first. Increased difficulty in dark and with eyes closed. May sway either side if bilateral involvement; if unilateral lesions then will sway to involved side. Patient watches his feet and keeps his eyes on the floor while walking.

- **Hemiplegic gait:** Arm and elbow flexed, wrist pronated, fist made with thumb inside, hip and leg extended, and foot plantar flexed. Patient walks with circumduction. Pelvis tilt is upward on the involved side. In foot drop high steppage may occur but foot will strike first; foot dragging may also occur.

- **Scissoring:** Knees may cross in front of each other. It is a bilateral hemiplegic gait. May occur with bilateral severe lesions as in congenital spastic diplegia, severe myelopathy.

- **Spastic ataxic gait:** Both spastic and ataxic components. May occur in vitamin B_{12} deficiency, MS, ALS.

- **Parkinsonian gait:** Stooped posture, short steps, head and neck forward, pin rolling tremors in hands, turns with whole body like a statue. Occurs in Parkinsonism.

- **Frontal lobe gait disorders:** Flexed posture, short shuffling steps, wide base, inability to integrate and coordinate lower-extremity movements. May occur with lesions of frontal lobe connections with the basal ganglia and the cerebellum. Gait apraxia (inability to walk in absence of motor weakness) may occur.

- **Marche a petit pas:** Short steps. No rigidity or bradykinesia as of parkinsonism. Occurs in old age, diffuse cereberal hemispheric dysfunction particularly of frontal lobe.

- **Gait in normal pressure hydrocephalus:** Short steps, wide base, gait apraxia.

- **Senile gait:** Slow speed, short steps, wide base.
- **Myopathic gait:** Exaggerated pelvic tilt (more than 5 degrees) bilaterally. Due to bilateral hip girdle weakness.
- **Trendelenburg gait:** Involved hip remains at level, and normal hip sags down during walking. Because of weakness of gluteus medius muscle due to lesion of superior gluteal nerve.
- **Hyperkinetic gait:** Dancing gait as in Syndenham's chorea.
- **Others:** Due to focal weakness of muscles like quadriceps weakness (femoral neuropathy).
- **Thalamic astasia:** Tendency to fall backward or to the side contralateral to the lesion.
- **Toppling gait:** Patient topples while walking. Due to cerebellar or brainstem lesion
- **Antalgic gait:** Painful gait.
- **Astasia-abasia** (Blocq's syndrome): Astasia (inability to stand), abasia (inability to walk). These occur in absence of normal motor functions while patient is recumbent. It is a conversion/psychogenic disorder.

Differential Diagnosis[3]

- **Infections:** These are the causes of acute causes of ataxia. There may be cerebellitis, brainstem encephalitis, most common agent is Varicella virus.[4] Others include EBV, influenza, hepatitis. Fever, rash generally accompanies.
- **Acute post-infectious demyelinating encephalomyelitis:** Along with the ataxia, there is alteration of consciousness with fulminant neurological deficits, seizures, and/or transverse myelitis. It develops during the recovery phase of viral illness.
- **Brainstem encephalitis:** Along with ataxia, there is long tract signs (hemiparesis), cranial neuropathies, respiratory irregularities, etc. Causes include EBV, Listeria, enteroviruses.
- **Drugs and Chemicals:**[5] Alcohol is the most common agent found. Anticonvulsants (like phenytoin, carbamazepine, phenobarbitone, vigabatrin, gabapentin, lamotrigine),[5] antihistamines (cimetidine), antineoplastics (5-FU, methotrexate, cisplatin, capecitabine), lithium, amiodarone, bismuth, isoniazid, metronidazole, glycoprotein IIb/IIIa

inhibitors, cocaine, heroin, chemicals (lead, manganese, aluminium, thallium, carbon-monoxide, insecticides, cyanide), hyperthermia are commonly implicated.

- **Space occupying lesions:** These present with slowly progressive ataxia and signs of raised ICP. Acute ataxia in these may be due to sudden hydrocephalous or haemorrhage inside the lesion. Posterior fossa masses are commonly associated, e.g. brainstem, cerebellar and cerbellopontine angle masses. Some supratentorial lesions like frontal lobe masses may also produce ataxia by involving fronto-cerebellar association fibres.

- **Trauma:** Cerebellar contusion or posterior fossa hematoma may produce ataxia. Vertebral artery dissection also may lead to ataxia due to ischemia of cerebellum.

- **Stroke:** Apart from vertebral artery dissection, thromboembolic disease may lead to stroke.

- **Sensory ataxia:** These may result due to involvement of posterior columns of spinal cord, spinal nerve roots, or peripheral nerves.

- **Motor (paretic) ataxia:** These may occur due to lesions anywhere in the motor pathway from motor cortex to the peripheral nerve.

- **Degenerative disorders:** These include Parkinsonism, Alzheimer's disease, dementia due to Lewy body disease, etc.

- **Functional:** These may occur in female adolescents.

REFERENCES

1. Campbell WW. *DeJong's THE Neurologic Examination*. Seventh. Philadelphia: Lippincott Williams & Wilkins; 2013.
2. Angel RW. The rebound phenomenon of Gordon Holmes. *Arch Neurol*. 1977;34(4):250.
3. Ryan MM, Engle EC. Topical review: acute ataxia in childhood. *J Child Neurol*. 2003;18(5):309–316.
4. Connolly AM, Dodson WE, Prensky AL, Rust RS. Course and outcome of acute cerebellar ataxia. *Ann Neurol*. 1994;35(6):673–679. doi:10.1002/ana.410350607.
5. Manto M. Toxic agents causing cerebellar ataxias. *Handb Clin Neurol*. 2012;103:201-213. doi:10.1016/B978-0-444-51892-7.00012-7.

Neurogenic Bladder

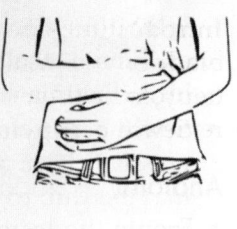

- Neurogenic bladder
  - History
    - Onset and duration progression
    - Frequency
      - Same
      - Increased
    - Hesitancy
    - Urgency
    - Voiding by suprapubic pressure/valsalva
    - Bladder sensations
    - Residual urine
    - Stream
    - Urinary retention
    - Social incontinence
    - Milky appearance
    - Sexual dysfunction
      - Erection
      - Ejaculation
    - Past history
  - Examination
    - Physical
      - Stream
      - Urinary retention
      - Social incontinence
      - Milky appearance
      - Sexual dysfunction
      - Past history
    - Neurologic
      - Sensory
        - Abdominal wall: T7-T12
        - External genitalia: S2,3,4
        - Groin: L1
        - Lower extremities: L2-S1
        - Perianal area: S2-S5
      - Motor
        - Lower limbs
        - Vol.contraction of anal sphincter
        - Anal tone
        - Neuromuscular reflexes

Introduction: The diagnosis and management of neurogenic bladder is a multidisciplinary task involving urological, neurological, and rehabilitation specialities. We intend to review the neurological part of this entity.

Anatomy of Bladder Connections

- Frontal lobe (paracentral Lobule): Centre for voluntary initiation of micturition. 99.8% of the time the lower urinary tract is in a phase of inhibition, and it is due to this centre.[1]
- Pons (Barrington's nucleus): Coordination. Results in inhibition of pelvic floor muscle activity during voiding.
- T9-L1 level Intermediolateral column: Gives sympathetic fibres which travel via sympathetic prevertebral plexus and hypogastric nerves to release noradrenaline predominantly on beta-adrenergic receptors at bladder dome, and on alpha-adrenergic receptors at internal urethral sphincter.[2]
- S2-S4 Intermediolateral column: It gives parasympathetic innervations by releasing acetyl choline at muscarinic receptors at bladder wall.
- Onuf's nucleus: It is present in the anterior horn cells of the S2-S4, and provides the somatic innervation of the striated external urethral sphincter via pudendal nerve.
- Afferent transmission of lower urinary tract stimuli travel via the pelvic, hypogastric, and pudendal nerves to the lumbosacral spinal cord.
- β-adrenergic stimulation causes inhibition of bladder contraction.
- α-adrenergic stimulation causes contraction of the bladder outlet. Thus sympathetic stimulation causes increased bladder storage.
- Micturition occurs by parasympathetic stimulation and reflex inhibition of somatic pathways to the urethral sphincter, and is controlled by pontine micturition centre.

Bradley's Four Loops of Control of Micturition

- Loop I: Frontal cortex, basal ganglia, thalamus and pons. Involved in inhibitory pathways and their inhibition causes partial or complete loss of volitional control of micturition reflex.

- Loop II: Pathway from detrusor muscle travelling in posterior column and lateral column up to brainstem and back in reticulospinal tract (no synapse in spinal cord). Partial interruption causes low threshold detrusor reflex. Complete interruption causes detrusor hyperreflexia. Incomplete voiding will result.
- Loop III: Detrusor and pudendal nerve motor nuclei and their interneurons in the spinal cord.
- Loop IV: From frontal lobe to pyramidal tracts to pudendal sphincter nucleus. Provides voluntary control of micturition.

History

- **Onset, duration, and progression**: Acute onset cases are generally because of trauma, infections, hemorrhage-involving frontal lobe or spine. Chronic progressive cases are because of malignancies, inflammations, multiple sclerosis.
- **Frequency**: Same or increased from routine. It is a sign of urinary tract infection. It also may be a sign of low volume bladder or detrusor irritability.
- **Hesitancy**: Difficulty in initiating micturition (detrusor-sphincter dyssynergia occurs in lesions above sacral cord, up to pons).
- **Sudden urgent micturition or urge incontinence**: Suprapontine lesions, incomplete suprasacral lesions. It is similar to infantile bladder.
- **Repeated bladder contraction** and voiding without a sensation of bladder distension or urgency (reflex bladder): It is because of complete suprasacral lesions.
- **Patient voids with abdominal straining or manual suprapubic pressure** (Crede manoeuvre) or Valsalva manoeuvre: It is seen in patients with sacral spinal cord lesions, peripheral denervation, and chronic urinary retention secondary to bladder outlet obstruction.
- **Paradoxical interruption** of the stream without voluntary control is consistent with detrusor sphincter dyssynergia.
- **Bladder sensations:** If absent, it indicates sensory neurogenic bladder (lesion at S2, 3, 4, cauda equina, conus lesion).
- **Residual urine:** Large residual urine indicates autonomous or sensory NB

- **Stream:** Poor indicates obstruction at lower urinary tract.
- **Urinary retention:** It may be seen in motor NB, e.g. in cauda equina syndrome.
- **Social incontinence:** It is because of lack of voluntary control of micturition, as seen in frontal lobe pathology.
- **Milky appearance of urine:** It is seen in retrograde ejaculation due to internal sphincter dysfunction.
- **Frequent uncontrollable precipitous voiding:** It is due to detrusor hyper-reflexia because of motor dysfunction of UMN type, e.g. in lesion above S2 level.
- **Sexual dysfunction:** Erection impaired in parasympathetic interruption, ejaculation impaired in sympathetic dysfunction.
- **Past history** of medical, neurological, urologic, obstetric, and gynaecologic problems or surgeries.

Examination

- **Bladder diary:** Patients should be told to maintain a bladder diary for a minimum period of two days.
- **Physical examination:** Abdominal examination, inspection of external genitalia, palpation of flank, digital rectal examination in men (prostate size, tenderness, consistency), vaginal examination in women (cystocele, mucocele, enterocele, uterine prolapse, malignancies), urethra (uretheritis, mass).
- **Neurourologic examination:** Sensory examination of the anterior abdominal wall (T7-T12), genitalia(S2-S4), and the lower extremities (L2-S1), perianal area (S2-S5).
- **Perianal sensations:** This can evaluate the afferent limb of pudendal nerve.
- **Motor examination** of lower limbs: Helps in localisation of the spinal level.
- **Voluntary contraction of the external anal sphincter** confirms innervations of the pelvic floor and integrity of corticospinal tract.
- **Preserved anal tone** in absence of voluntary anal contraction indicates suprasacral lesion, while diminished anal tone indicates a sacral or peripheral nerve abnormality.
- **Neuromuscular reflex:** Bulbocavernosus, anal, and cremasteric.

- **Bulbocavernosus**: Afferent limb: Pudendal nerve, cord level: S2-4, Efferent limb: Pudendal nerve.
- **Anal:** Afferent limb: Pudendal nerve, Cord level: S2-S4; Efferent limb: Pudendal nerve.
- **Cremasteric:** Afferent limb: Femoral branch of genito-femoral nerve; Cord segment: L1,2; Efferent limb: Genital branch of genitofemoral nerve.

Investigations

1. **Urine routine microscopy and culture:** These are ordered if infection is suspected.
2. **USG abdomen:** It is ordered to see for post-void residual volume, prostate volume, bladder wall changes, and upper urinary tract changes.
3. **MRI spine/brain:** It is ordered based on the suspected spinal or cortical lesion.
4. **Urodynamic studies:** It is the investigation of choice for knowing the functions of detrusor muscle and sphincters. Videocystometry can tell both functional and structural changes of bladder, and uses fluoroscopic monitoring after filling of the bladder with contrast agent.
5. **Pelvic neurophysiology tests:** EMG of anal and bladder sphincters can be ordered if sacral cord or peripheral nerve lesions are suspected. These are less used now, after the introduction of videocystometry.

Important Points

- Lesions below L1 vertebral body leads to injury to the cauda equina, and clinical pattern suggestive of peripheral dener-vation.
- From spinal cord: Sympathetic fibres → sympathetic chain ganglia → prevertebral ganglia in the superior hypogastric and pelvic plexus → bladder and urethra.
- From spinal cord: Parasympathetic fibres travel through sacral spinal nerves → form pelvic splanchnic nerves → coarse with sympathetic nerves in the pelvic plexus → bladder wall
- In infants bladder functions are purely reflex because of lack of complete myelination (similar to frontal lobe pathology).

- Mass reflex: Pathological exaggeration of sexual reflex. Priapism and ejaculation occurs with minimal stimulation
- Root value of pudendal nerve is S2-S4.
- **Level of lesions and presentation:[3]**
  - Distal to sacral cord (Cauda Equina): There will be loss of innervation to the bladder, and presentation may vary depending upon the loss of sensory, motor or complete control.
  - Below T6 and above S2-S4: Detrusor overactivity, absent sensation below the area of the lesion, smooth sphincter synergy, striated sphincter dyssynergy. Due to this, during voiding there will be simultaneous contraction of detrusor and sphincters leading to incomplete bladder emptying and high intravesical pressure.
  - Above T6 up to pons: Also includes smooth sphincter dyssynergy.
  - Above pons: Bladder is like that of an infant. There will be no volitional control of micturition, either to start or control micturition, due to loss of control by frontal lobe over Barrington's nucleus of pons. Patient will have detrusor overactivity, and will have urge incontinence.
- Spinal shock: After spinal injury, initially there is flaccid motor paralysis, and fixed bladder neck (competent and closed) leading to retention with overflow incontinence. After recovery of spinal shock, detrusor sphincter dyssynergy ensues.
- Autonomic dysreflexia (pounding headache, hypertension, bradycardia, and flushing with sweating) above the zone of lesion. It occurs with lesions above T6. More common with cervical lesions. Provocative stimulus is generally bladder distension or manipulation of the bladder or rectum.
- Neurodysraphism: Classic description: Areflexic bladder with open bladder neck.

  Other more common: Detrusor overactivity, or poorly compliant bladders, fixed external urethral sphincter, with 10–15% patients having detrusor striated sphincter dyssynergia. Patients may often have overflow and stress incontinence.

Table 24.1: Classification of neurogenic bladder (based on Urodynamic criteria): Lapipdes

Type of NB	Mechanism	Voluntary control	Sensations	Residual urine	Bladder Capacity
Unin-hibited	Loss of cortical inhibition	Absent	Present	No	Vary
Reflex	Severe myelopathy, extensive brain lesions, micturition is reflex	Absent	Present	Variable	Small
Autono-mous	No external innervations to bladder. Micturition is by intrinsic neural plexus	Absent	Absent	Large	Not greatly incre-ased
Sensory paralytic	Lesions of posterior root ganglia	Present, but hesitancy, dribbling is present, difficulty in emp-tying bladder	Absent	Large	Increases
Motor paralytic	Motor nerve supply to the bladder is cut	Absent, bladder distends and decom-pensates	Normal	Vary	Vary

Differential Diagnosis

1. **Multiple sclerosis:** Lower urinary tract symptoms (LUTS) are present in 75–80% of patients, urinary incontinence in more than 50% of patients, and depend upon the duration and severity of the disease.[4,5] Along with the higher mental function disturbances, the presentation will be loss of voluntary initiation and control of micturition due to loss of cortical control, leading to urge incontinence and double voiding. Urodynamic study will show findings of detrusor overactivity, detrusor hypocontractility, and detrusor-sphincter dyssynergy.

2. **Parkinson's disease (PD):** LUTS presents in 27–63.9% of patients with incontinence in 33% of them.[6] In PD, there is loss of dopaminergic output resulting in loss of tonic inhibition, that results into detrusor overactivity. Frequency, urgency and difficulty voiding may ensue.

3. **Multiple system atrophy:** There is neuronal degeneration at multiple sites leading to varying presentation. There may be detrusor overactivity, overflow incontinence, or stress incontinence.

4. **Stroke:** Urinary symptoms appear more when the lesion involves mesial frontal regions, capsule, putamen, and thalamus. Detrusor overactivity occurs more than detrusor underactivity. Urge incontinence occurs because of them.

5. **Dementia:** Symptoms are same as the above mentioned cortical diseases. Both NPH and Alzheimer's disease may present with incontinence, though more common in NPH.

6. **Frontal lobe masses:** Urinary features are same as of the above mentioned cortical diseases. Features of raised ICP, and behavioural distubances may also be found.

7. **Spinal cord injury (SCI):** Urinary symptoms are found in about 85% of the SCI patients, with the incontinence reported in about 52% of patients.[6] These vary according to the level of injury, and are explained above.

8. **Spina bifida:** Urinary symptoms are present in about 90% of patients. Urodynamic study may show features of detrusor overactivity, underactivity, or detrusor-sphincter dyssynergy.

9. **Disc prolapse:** They may present with UMN type or LMN type bladder depending upon the level. Other features like radicular pain, foot drop, etc. may be elicited in the history.

10. **Inflammations (tuberculosis):** They may also present with UMN or LMN type bladder depending upon the level involved. Other features like low grade fever, tenderness, neurological deficit, appetite and weight loss may be seen.

11. **Spinal tumours:** Both primary and secondary tumours may present with UMN or LMN type bladder depending upon the level involved. Other features like tenderness, back pain more at night, neurological deficit, appetite and weight loss may be seen.

12. **Autonomic disturbances:** Autoimmune autonomic ganglionopathy, acute idiopathic autonomic neuropathy, pure autonomic failure, postural tachycardia syndrome are the syndromes in which urinary disturbances may occur. Features are variable.

13. **Peripheral neuropathy:** Diabetes mellitus, Guillain-Barré syndrome may cause urinary symptoms by affecting the peripheral nerves. Both detrusor areflexia and overactivity may occur.

REFERENCES

1. Panicker JN, Fowler CJ. The bare essentials Uro-Neurology. *Pract Neurol*. 2010;10(3):178–185.

2. Andersson K-E, Arner A. Urinary bladder contraction and relaxation: physiology and pathophysiology. *Physiol Rev*. 2004;84(3):935–986.

3. Winn HR. *Youmans and Winn Neurological Surgery*. 6th ed. New York: Elsevier

4. De Sèze M, Ruffion A, Denys P, Joseph P-A, Perrouin-Verbe B. The neurogenic bladder in multiple sclerosis: review of the literature and proposal of management guidelines. *Mult Scler J*. 2007;13(7): 915–928.

5. Fowler CJ, Panicker JN, Drake M, et al. A UK consensus on the management of the bladder in multiple sclerosis. *J Neurol Neurosurg Psychiatry*. 2009;80(5):470–477.

6. Ruffion A, Castro-Diaz D, Patel H, et al. Systematic review of the epidemiology of urinary incontinence and detrusor overactivity among patients with neurogenic overactive bladder. *Neuroepidemiology*. 2013;41(3-4):146–155.

Index

3 Oz test   137

A

Accommodation reflex   55
Alexander's law   132
Anosmia   44
Aphasia   37
Apraclonidine test   71
Articulation   35
Astasia-abasia   208
Ataxia   187, 199
Athetosis   187
Atypical facial pain   82
Aura   27

B

Ballism   187
Barrington's nucleus   211
Bell's palsy   91
Bergara-Wartenberg sign   94
Blocq's syndrome   208
Bradley's four loops   211
Bragard's sign   152
Brainstem auditory evoked
    response   110
Broca's aphasia   12, 35, 37, 38
Bruns nystagmus   127
Burning mouth syndrome   88

C

CALFRAST   109
Caloric reflex test   122
Carpal tunnel syndrome   143
Cerebral ptosis   73
Cervical distraction test   145
Cervical radiculopathy   141, 146
Choosing Wisely Campaign   14
Chorea   187
Cluster headache   19

Cohen syndrome   53
Comprehension   36
Conduction aphasia   37
Confrontation test   55
Corneal reflex   85
Cover/uncover tests   64
Crack pot sign   5
Cyclic vomiting syndrome (CVS),
    41, 43

D

Deep tendon reflexes   179
Dementia   7, 138
Diplopia   60
Doll's eye reflex   123
Duane's syndrome   69, 72
Dysarthria   34–36, 38
Dysdiadochokinesia   205
Dysmetria   204
Dysphagia   133
Dysphonia   36
Dyssynergia   204
Dystonia   187, 198

E

Epilepsy   24
Electro-encephalogram (EEG)   30
Endurance   167
Epley's manoeuvre   121

F

FABER test   152
Facial asymmetry   90
Facial pain   81
Fascioscapulohumeral
    dystrophy   98
Femoral stretch or Ely test   152
Foster-Kennedy syndrome   49
Fresnel goggles   122
Fukuda's stepping test   124

G

Gag reflex 137
Gait ataxia 201
Grating acuity 54

H

Head circumference 5
Headache 1, 2, 18
Hearing loss 106
Hemeralopia 52
Hemifacial spasm 97
Hertel's ophthalmometer 77
Higher mental function
    examination 8
Horner's syndrome 69
House Brackmann grading 96
Hydroxyamphetamine test 71
Hyperacusis 93

I

ICP monitoring 17
Ictal nystagmus 27

J

Jaw jerk 85

L

Lacrimatory pathway 92
Lasègue test 152
Lhermitte's sign 145, 159, 163
Low back pain 149
Lumbar puncture 16
Luria test 10

M

Macewen sign 5
Magnetoencephalogram
    (MEG) 31
Marcus Gunn jaw-winking
    syndrome 69, 72
Mendelson manoeuvre, 135
Migraine 2, 3, 5, 18, 21, 125
Miller Fisher syndrome 73
Minimental state examination 6
Hindi minimental state
    examination 7
Movement disorders 186
MRI brain (epilepsy protocol) 30

M

Mutism 35
Myoclonus 187

N

Neck pain 141
Nyctalopia 52
Nystagmus 126

O

Oculocephalic reflex 123
Onuf's nucleus 211
Ophthalmoplegic migraine 73
Ophthalmoscopy 56
Optic atrophy 56
Optotype acuity 53

P

Papilledema 52
Paracusis Willisi 107
Parkinson's disease 187, 197, 217
Parosmia 45
Past pointing 123
Pathological reflexes 180
Phantosmia 45
Physiological nystagmus 131
Postural orthostatic tachycardia
    syndrome (POTS) 42
Projectile vomiting 41
Proptosis 75
Pseudonystagmus 131
Pseudoptosis 73
Ptosis 67
Pupillary reflex 55

R

Raised ICP 19
Ramsay Hunt syndrome 93, 98,
    103
Rebound phenomenon 205
Recruitment phenomenon 107
Retropulsion pull test 192
Rigidity 191
Rinne test 110
Romberg's test 123

S

Schirmer's test 95
Seizures 4, 24, 25, 27

Sensory impairment   157
Sicard's sign   152
Sniff magnitude test (SMT)   47
Sniffin' sticks   47
SNOOP   14
Spasmodic torticollis   147
Spasmus nutans   126
Spasticity   191
Speech   34
Speech discrimination   108
Spurling's sign   145
Star walking test   123
Straight leg raising test   152
SUNCT   87
Superficial reflexes   179

T

Taste   100
Taste pathway   92
Tensilon test   65
Tension headache   19
Tests of olfaction   47
Thalamic astasia   208
Tics   187

Tinnitus   113
Todd's paralysis   27
Tone decay   108, 111
Tremor   187, 194
Trigeminal neuralgia   82, 86
Tuning fork tests   109

U

University of Pennsylvania
       smell identification test
       (UPSIT) 47
UPDRS score   192

V

Vernier acuity   54
Vertigo   119
Videoflouroscopic swallow study,
       138
Visual disturbance   51
Vomiting   40

W

Weakness   165
Weber test   110
Wernicke's aphasia   37